Galileo's Muse

Galileo's Muse

Renaissance Mathematics and the Arts

Mark A. Peterson

Harvard University Press
Cambridge, Massachusetts
London, England
2011

Library of Congress Cataloging-in-Publication Data
Peterson, Mark A., 1946–
    Galileo's muse : Renaissance mathematics and the arts / Mark A. Peterson.
        p.   cm.
    Includes bibliographical references and index.
    ISBN 978-0-674-05972-6 (alk. paper)
        I. Galilei, Galileo, 1564–1642.   2. Arts, Renaissance—Italy.
3. Mathematics—Italy—History.   4. Science and the arts—Italy—History.
I. Title.
    QB36.G2P48   2011
    709.02′4—dc23   2011023319

Contents

Contents

Galileo's Muse

Prologue

Let me not look like a flatfish on a silver platter.

With something like these words, a mysterious author, to be encountered and perhaps unmasked in the last chapter, contemplates how foolish he will look when it turns out that his ambitions outstrip his abilities. He hesitates to attempt something that prudence warns him is risky, yet "though you drive it away with a pitchfork," he says, quoting Horace, "it keeps coming back."

I know how he feels, introducing a book that might well be called overambitious. I can only say that it didn't start out that way. The beginnings of this project were some observations about mathematics and the arts in the late Middle Ages and early Renaissance. I was especially intrigued by some mathematical ideas that I had noticed in Dante, unexpected mathematical sophistication centuries before Galileo. I also became fascinated with Galileo, and I began to wonder where he had come from. This question seemed to organize my thoughts. What was Galileo's intellectual inheritance, and how did it form him? Galileo's education was in the humanities and the arts, so the question is a sprawling one. And even that is not enough, because Galileo's ultimate enthusiasm was for mathematics, and that is another broad intellectual stream. Where all these streams mixed, that is where

I

Galileo came from, or so I imagined. To understand it, I had to follow the streams back to their sources.

The structure of this book is largely determined by the process of following one stream after another. After an introductory chapter on Galileo, establishing the necessity of looking back to classical sources, I summarize the classical legacy in mathematics and the sciences. In particular I emphasize a distinction between Greek and Roman contributions to this legacy that turns out to be crucial for understanding Galileo and Renaissance mathematics more generally. There follow four sections, each containing two chapters, on several Renaissance arts: poetry, painting, music, and architecture. Each section addresses the attempt to recover classical excellence. Perhaps unexpectedly there is a clear mathematical thread running through every one. Implicitly I ask, what did mathematics mean for the arts? And what did the arts mean for mathematics? These chapters are largely self-contained and could be read in any order. A chapter on Renaissance mathematics itself brings me back to a penultimate chapter on Galileo, now considered in light of all this background material.

That might have been the end of the book, but as I was attempting to tie up some loose ends, I made a discovery, a previously unrecognized work of Galileo. Not everyone will agree with this attribution, of course, but I don't even ask for such agreement, necessarily. The mere existence of this little book and the way it came to be published pushes us to look at Galileo in the way that I have set out to do here. It represents his views, beyond question, and I take it as confirmation. *Now, where is that silver platter?*

1 Galileo, Humanist

Following his condemnation by the Church in 1633, Galileo's long and remarkable career should have been finished, but old as he was, nearly blind, under house arrest, and forbidden to publish, his last and ultimately most influential book suddenly appeared in print in 1638. My, my, what a surprise, he says, in effect, in the dedication.[1] What have we here? The book was *Two New Sciences*, a rambling dialogue full of highly original ideas and discoveries, including especially the famous parabola law: "It has been observed that missiles and projectiles describe a curved path of some sort; however no one has pointed out the fact that this path is a parabola. But this and other facts, not few in number or less worth knowing, I have succeeded in proving; and what I consider more important, there have been opened up to this vast and most excellent science, of which my work is merely the beginning, ways and means by which other minds more acute than mine will explore its remote corners." Galileo's suggestion that there would be a vast territory to explore was justified. Fifty years later Isaac Newton would cite Galileo's parabola law as the first and simplest illustration of the laws of motion in his *Principia Mathematica*, the beginning of Newtonian mechanics and all of modern physics, confirming Galileo's claim to vastness beyond what anyone could have foreseen.

Thus it is not just in hindsight that we see Galileo as a transitional figure, living at the epicenter of a scientific revolution. He saw himself that way too. Historians of science have focused on this moment as if it were the key to understanding the scientific revolution itself. As R. H. Naylor put it, "No one doubts that Galileo played a major part in changing the European view of physical science. Looking at his *Two New Sciences* (1638) and comparing it with the work of his immediate predecessors and contemporaries we sense the existence of an immense gulf. This radical break becomes all the more striking when, on looking closer we discover its presence in his own work. How it came about is one of the great unsolved mysteries of European intellectual history."[2]

Scholarship on the parabola law and its origins has centered, appropriately, on the development of theories of motion, especially the medieval "impetus theory." Galileo knew this tradition and even worked within it in his initial efforts to understand motion, but his final formulation in *Two New Sciences*, although it still contains certain formal features derived from impetus theory, doesn't mention the medieval tradition at all. What he says sounds almost misleading: "I wonder not a little how such a question escaped the attention of Archimedes, Apollonius, Euclid, and so many other mathematicians and illustrious philosophers . . . There is a fragment of Euclid which treats of motion, but in it there is no indication that he ever began to investigate the property of acceleration."[3] That is, Galileo ignores two thousand years of history and connects his work directly to the Hellenistic Greeks instead, even if there is no real precedent to point to. It is no wonder if this remark has been considered unhelpful in understanding what Galileo did.

Historians are accustomed to a certain caginess on Galileo's part when it comes to the attribution of ideas and motives. His mock surprise at seeing his book about to be printed, already noted in the dedication to *Two New Sciences*, is a transparent example that one could multiply many times.[4] Still, when Galileo connects the parabola law to the Hellenistic Greeks, he seems to mean it genuinely. The telltale tongue-in-cheek tone is missing, and he sounds sincere. *Parabola* is a Greek word, after all. The Greeks had known everything necessary to discover the parabola law, even if they hadn't actually discovered it. The simplest interpretation of these lines, reinforced in many other places, is that Galileo really did regard himself as the direct heir of the Greeks. Most of this book will explore what that means.

It is tempting to argue that Galileo's humanism predisposed him to honor what was classical and pass over in silence what was medieval. It would be too facile, though, to assume that Galileo therefore failed to recognize or acknowledge the real source of his own work. As historians like Naylor, above, have pointed out, Galileo's work is not just an incremental improvement on previously existing theories. Something mysterious happened, and it makes sense to look for it where Galileo himself directs us, in the classical past, even though the search must be indirect and may have nothing immediately to do with theories of motion at all.

All Galileo biographies tell us that Galileo's education was in the humanities and the arts, and occasional scholarly essays even urge a fuller exploration of what that might mean.[5] An early example, and perhaps the most interesting, is offered by the cultural historian and critic Erwin Panofsky in a little book called *Galileo as a Critic of the Arts*.[6]

Panofsky presents an unfamiliar Galileo, seen from his humanistic side. He discusses, among other things, a letter that Galileo wrote to the painter Cigoli, defending the superiority of painting over sculpture, apparently in response to his friend Cigoli's request for some good arguments on this point. There are many things that a modern reader might find surprising here. First, that Cigoli, one of the most celebrated artists of his day, would make such a request of Galileo; second, that Galileo would respond the same day, and with complete engagement, to a question that might seem to have nothing to do with him;

and third, that he would propose amusing and novel arguments in a debate that had been going on, as an intellectual game, for decades if not centuries. Panofsky rightly points out parallels in this little incident to Galileo's scientific work: "If Galileo's scientific attitude is held to have influenced his aesthetic judgment, his aesthetic attitude may just as well be held to have influenced his scientific convictions. To be more precise, both as a scientist and as a critic of the arts he may be said to have obeyed the same controlling tendencies."

Panofsky did not pursue this tantalizing point much beyond the level of a casual suggestion, and he never dealt with Galileo's mathematics at all. Galileo's mathematics is just as essential to his humanism as is his erudition in the arts, but studying these things in combination requires more kinds of expertise than perhaps anyone can honestly claim. Thus the significance of Galileo's humanism for his scientific thought, while often acknowledged, remains largely unexplored. And this is true even while unresolved questions about the nature of Galileo's scientific thought persist.

In writing this book, I am not claiming special expertise. Just the opposite. I have a generalist's role, assembling evidence and synthesizing a narrative that makes sense of the interaction of mathematics and the arts in the Renaissance. That Galileo drew upon mathematical traditions in the arts in his scientific work is the motivation, and sometimes the substance, of this narrative. I will move from Galileo himself to the many legacies that he inherited, from classical mathematics to traditions in the arts, aiming to build a picture of Galileo's intellectual world, the world of his discoveries.

The View from Arcetri

Galileo lived out his last years under house arrest in his little villa Arcetri just outside of Florence. He was allowed visitors, although his own movement was restricted, and admirers from all over Europe did visit him, including the young John Milton and the middle-aged Thomas Hobbes. Galileo's sovereign, Grand Duke Ferdinand II de Medici, condescended to pay him a long visit in 1638, when he was seventy-four years old. From late 1639 until his death in January 1642, a promising young man named Vincenzo Viviani lived with him in his house as his student and amanuensis. And he had neighbors.

There are two surviving biographical essays written by people who knew Galileo in those days, that of his student Viviani and that of a neighbor, Niccolò Gherardini. There are also some short biographical notes by his son, Vincenzo Galilei.[7] All three essays remained unpublished until the eighteenth century. They are all roughly datable to the 1650s, ten years or so after Galileo's death, although Viviani continued to work on his throughout his long life and only failed to publish it because he never felt that it was quite equal to its great subject.

Viviani collected all kinds of documentary evidence for this project, and it is thanks to him that the Galileo manuscript collection, now in the National Library in Florence, is as comprehensive as it is. In spite of all his documentation and his sincere attempt to be accurate, Viviani is not always correct in matters of fact. The other two, working from memory, are completely unreliable. Gherardini cheerfully admits that he cannot remember the names of important personages in Galileo's life. For the period of his old age, though, these biographies are themselves primary documents because their authors knew and talked with Galileo.

Each of these little biographies is independent of the others, as is clear from their many differences, even when describing the same events, and each author reveals, with a little reading between the lines, his own distinct relationship with Galileo. Gherardini notes with appreciation, for example, that Galileo did not at all insist on talking about mathematics or science (Gherardini was trained in the law), but that he would adapt his conversation to whomever he was with, and that he was so fond of company that he could hardly bear to eat alone. Gherardini had first met Galileo only in 1633 in Rome, at the time of Galileo's trial. That he felt able to write a biography solely on the basis of conversations with Galileo afterward, while understanding nothing of Galileo's scientific work, tells us that Galileo must have done a lot of reminiscing. In addition, Gherardini and Galileo seem to have enjoyed discussing philosophical topics.

Viviani's biography is the longest of the three, and is the source for several well-known stories about Galileo, like the public demonstration of falling cannonballs at the Leaning Tower of Pisa and the constant period of the swinging lamp in the cathedral. The single largest topic in Viviani's essay is Galileo's long, ultimately unsuccessful effort to solve the longitude problem by using the moons of Jupiter as a clock, a project in which even Viviani himself had been involved at the end. His deep affection and reverence for Galileo is evident on every page.

Galileo's son Vincenzo seems not to have known his father very well. His biographical sketch is only three printed pages long, and a surprisingly large portion of it is devoted to Galileo's last years, after the trial, when Vincenzo did spend time with him. He emphasizes the high respect that his father still commanded, even in his old age. He describes some of the scientific questions on which Galileo was working in his last years and regrets that so many philosophical and mathematical propositions were still not written down when he died, a great loss.

The rough outline of the events of Galileo's life can be extracted from these documents, even if subsequent biographers have had to make major additions and corrections. Born in 1564, the son of Vincenzo Galilei, musician, music theorist, and occasional wool merchant, Galileo became professor of mathematics at the University of Pisa in 1589 and then at the University of Padua in 1592. In 1609, hearing of the existence of a telescope, he was able to make one, and then improve it, to such a degree that within just a few months he made a series of spectacular astronomical discoveries. These included the mountains on the Moon, the resolution of the Milky Way into faint stars, and the four moons of Jupiter, christened by him the Medicean Planets in the book that he quickly published, dedicated to the Medicean grand duke of Florence, Cosimo II. His cultivation of the Medici led to his appointment in 1610 as First Mathematician and Philosopher to the grand duke, with a high salary and virtually no official duties. This enviable position was indeed envied, and jealousies and controversies became part of his life, as he was attacked on various fronts. Most famously his enemies sought to tie him to Copernicanism, the heliocentric model of the universe, and to bring Copernicanism into disrepute, succeeding very well on both counts. All modern biographies spend a lot of time on something that the first biographies hardly mention, Galileo's 1616 trip to Rome,[8] trying to forestall what then happened, the declaration by the Church that the motion of the Earth was absurd in philosophy and contrary to Scripture. On this occasion Galileo was served with an injunction not to hold, teach, or defend the Copernican doctrine, although whether the injunction was legal, and what the exact intention had been, were essential questions in the trial seventeen years later. Believing he had obtained permission in 1624 to write a book comparing the Sun-centered Copernican system with the Earth-centered Ptolemaic system, he published his great *Dialogue on the Two Principal World Systems* in 1632, only to see it immediately banned and himself accused of violating the 1616 injunction. By the end of the trial, even Galileo had to agree that he had made too good a case for Copernicanism, and he lived under the Church's sentence for the rest of his life. He was detained for a few months at the palace of his friend Archbishop Piccolomini of Siena, who did his best to make up for the indignity Galileo had suffered, encouraging him to write what eventually became *Two New Sciences,* and in early 1634 he was allowed to return to Arcetri in Florence.

Despite his humiliation in Rome, despite the physical ailments that plagued him, and despite most pathetically the death, soon after his return, of his beloved daughter Sister Maria Celeste, which he took very hard, Galileo's life at Arcetri gradually became almost happy. He had leisure to think about his life and work as a whole. In scientific terms, that is what *Two New Sciences* is. Its problems are the problems that occupied him all his life, from the days of his first professorship at Pisa, and even before that. In writing it, he was summarizing his life's work. He even included in an appendix some lovely theorems in the style of Archimedes that he had proved as a young man more than fifty years earlier. The book is what he had intended his work to be before the Copernican controversy interrupted and distracted him.

Education

Galileo's first biographers—his son, his neighbor, and his student—have quite a lot to say about this period of his retirement, even if not much was happening in the way of grand biographical events, and they indirectly record their own lives along with his. Two of them mention Galileo's gardening, for example. He must have offered his guests fruit or wine of his own cultivation. The only one who doesn't mention gardening is Viviani, who, to judge by his choice of topics, was completely absorbed in the scientific work he was doing with Galileo. Despite the marked differences among them and their inconsistency when they are talking about remote events, all three together paint a consistent picture of life at Arcetri.

There are interesting differences, though, in what they report Galileo saying. As Gherardini had noted, he adapted his conversation to whomever he was talking with, and this adaptation sometimes extended to what one might consider matters of fact. If he meant to instruct by means of a story, say, then the important thing would be the lesson imparted, not the literal truth of the episode. As he said in another context, in works like *The Iliad* or *Orlando Furioso* "the least important thing is whether what is written there is true."[9] He did not say this in any way to denigrate *Orlando Furioso*, his favorite poem, a chivalric fantasy, but only to point out that poems and stories have other purposes. We can assume that this principle would also apply to his own stories. It is an additional layer of complexity that we encounter his stories at second hand in works of biography that also aim to instruct, or at least to memorialize.[10] But there can be no mistaking that some of the inconsistencies must have originated with him. In light of Galileo's continual refashioning of himself, to use Mario Biagioli's phrase,[11] one should not be surprised if Galileo's idea of what is true sometimes appears rather flexible. Vincenzo says, "No vice was more detestable to him than dishonesty, perhaps because in the mathematical sciences he recognized all too well the beauty of the truth,"[12] and Vincenzo probably meant that straightforwardly, but it is more complex than that. Even in the mathematical sciences the truth is not so simple, something Vincenzo didn't have the personal experience to appreciate. If Galileo had had too rigid a conception of the truth, he could not have been Galileo.

As he thought and wrote about his scientific work, it is clear that he also reminisced about his own childhood and his education, because all three biographers include such stories. With Viviani he used the stories didactically. He emphasized his family's poverty, saying that his father wanted to send him away to school, but couldn't afford it. Instead, he says, he excelled by his own perseverance and hard work, in spite of mediocre teachers. Gherardini corroborates this. In fact, though, Galileo had spent two years at boarding school at a monastery in nearby Vallombrosa. By omitting this detail Galileo made his own education into a parable of self-reliance and self-discipline for the benefit of Viviani, who was both poor and deserving, being supported by a scholarship from the grand duke.

We cannot be sure, in fact, that Galileo did not have good teachers.

All that Viviani knew about the Vallombrosan connection is that Galileo had studied Aristotelian logic with a Vallombrosan friar, but had found it tedious, or, as Viviani put it, "of little satisfaction to his exquisite intellect." Here too Galileo seems to be turning the episode into a lesson for Viviani, because he emphasizes the importance of going beyond mere memorization, by rote, of "dialectical terms, definitions, and distinctions," his description of his lessons in logic. It is worth noting, though, that elsewhere Galileo is grateful that he can use Aristotelian logic with confidence,[13] something that, according to this testimony, he would perhaps not have had the patience to learn on his own. He might have had good teachers after all.

Young Galileo's first subject was Latin, the basis of a humanist education. Both Gherardini and Viviani relate that he rapidly outstripped his teacher's ability to help him, and taught himself by reading the best classical authors. There is no reason to doubt this. It is clear on other grounds that he cultivated a good writing style from an early age. Galileo's Latin prose style, when he needed to call upon it, was adequate to his professorial duties, and also to that charmed moment in 1610 when he described his new telescopic discoveries to a European audience in the *Sidereus Nuncius* and suddenly became Europe's greatest astronomer. Key passages in *Two New Sciences* are also in the scientific language of Latin, as are some early unpublished attempts at a theory of accelerated motion. If his other published works and most of his correspondence are in Italian, it cannot be because of any deficiency in his Latin.

Galileo wrote for a general audience, one clear reason for preferring Italian, but his sheer love of the Italian language is undoubtedly another, even when, as in the *Dialogue on the Two Principal World Systems*, he claimed to be addressing transalpine readers.[14] According to Gherardini he was proud of making the Tuscan language, through his own work, a scientific language,[15] in the same way (although he doesn't say this) that Dante had made the Tuscan language a literary language. He worked hard at writing and credited his graceful Italian style to the careful study of Ludovico Ariosto, his favorite poet. Galileo's concern for good style is almost visceral. His copy of Ariosto is heavily annotated with his own comments on the poet's choices, and he systematically compared the styles of Tasso and Ariosto almost line by line, to the great detriment of Tasso. Although he had made these notes in his youth,[16] he was still thinking about them in his old age: he told Gherardini that reading Tasso after Ariosto was like eating sour little citrons after delicious melons. He very much regretted the loss of those old notes, Gherardini says, because he had intended them to be seen and read. (They have since turned up again,[17] and they are very entertaining, often scurrilously funny at Tasso's expense.) Gherardini notes that in conversation Galileo would freely quote Ariosto to make a point in a memorable way. One can find occasional examples of this in Galileo's publications as well.

According to Gherardini, "He was most expert in all the sciences and the arts, as if he were professor of them. He took extraordinary delight in music, painting and poetry." Galileo's accomplishment in the arts could only have come from his early education, as Viviani tells us in more detail. He learned to play the lute from his father, and, Viviani says, surpassed him in the sweetness and grace of his playing, maintaining this ability to the end of his days. He learned drawing, in which he was marvelously gifted. The many excellent painters who were among his dearest friends respected his talent, and even deferred to him in questions of taste and design.

Again, there is no reason to doubt this. All the biographers agree that Galileo played the lute, and if Viviani says that he still played in his old age, then he must actually have heard him. As for Galileo's ability and taste in drawing, there is ample confirmation elsewhere. Galileo was indeed an excellent draftsman, and the illustrations that he made for his own books are still occasionally pointed out as little masterpieces of visual argument.[18] It is true that the painter Cigoli was one of

Galileo's best friends around the time of his telescopic discoveries, as Viviani says. Cigoli wrote often to Galileo, keeping him informed about the art scene in Rome, once even passing along a literary request.[19] Artemisia Gentileschi turned to him, as to an old friend, when she felt that one of her paintings had not been sufficiently noticed by the grand duke.[20]

An education in the humanities was preparation for professional study, although Galileo told Viviani that if he could have chosen his profession in those years, he would have been a painter. Galileo's father sent him to the University of Pisa at the age of seventeen to earn a medical degree and restore the family's fortunes. This plan, however, did not turn out well. Viviani describes Galileo's encounter with the curriculum there in a tone of horror, that Galileo, "who was chosen by nature to unveil to the world secrets of nature that for ages had lain hidden in the densest obscurity" should "give himself up . . . almost blindly" to the assertions of others. As a "free intellect, he didn't think he should have to assent so readily to the sayings and opinions of ancient and modern writers, when he could serve himself better with reasoning and the experience of the senses."[21]

Galileo's confidence in reasoning and the experience of the senses must have developed and grown over the course of his career, but for Viviani's benefit he places it right at the beginning. A related theme had occurred even earlier, in Viviani's account of Galileo's childhood:

> In his play he would endeavor to make with his own hands little devices and machines, imitating with miniature models contrivances that he had seen, such as mills, galleys, and every other commonplace invention. And when he lacked some necessary part for one of his boyish constructions, he would invent a replacement, using pieces of whalebone in place of iron springs, or something else for some other part, according as the need suggested, adapting to the machine new ideas and tricks of mo-

tion, so that it would not be left unfinished and so that he could see it operate.[22]

Viviani and Galileo did not do experiments together, and what is described here is not experimentation, but in this story Galileo seems to want to call attention to the inventiveness that characterizes a good experimenter, able to fashion replacement parts and quick to tinker. Interestingly, young Isaac Newton is also said to have made toy mills. Whatever the truth of it, Galileo wanted to be thought of as a boy who would have done this. And as for his inventiveness in his later career, there can be no question about it. His rapid invention of a working telescope, based on the most perfunctory description of such a device, followed within weeks by greatly improved lenses of his own grinding, is only the most obvious example.

Nothing in the early years suggests what was about to happen, an event that changed Galileo's life, his discovery of Euclidean geometry at about the age of nineteen while he was chafing at his studies in medical school. Euclidean geometry was not part of the program of study, and how he came to it is different in each version. In Viviani's telling, he came to mathematics through the arts. Young Galileo had never really thought about mathematics, Viviani says, "but the great talent and delight that he had, as I said, in painting, perspective, and music, hearing his father frequently say that such things had their origin in geometry, moved in him a desire to try it, and he asked his father many times to introduce him to it."[23] His father refused, not wanting to distract him from his medical studies, but Galileo appealed to a family friend, Ostilio Ricci, a good mathematician, to teach him Euclid, and he immersed himself so thoroughly in the subject that his father forbade him to continue. Galileo had to hide the Euclid under his medical books so that he could pretend to be studying medicine if his father should come in. Eventually his father recognized his mathematical talent and consented to his chosen career.

Gherardini's version is roughly consistent with this but more conspiratorial, in keeping with Gherardini's own predilections.[24] In this version Galileo accidentally overhears a geometry lecture by Ricci to the court pages and goes on to listen clandestinely for weeks before finally revealing himself to his unsuspecting teacher. Gherardini confirms that Galileo's father opposed his study of mathematics and that Galileo deceived him by keeping his Euclid under the medical books and so forth, in what is essentially an even more elaborate version of Viviani's story.

Vincenzo Galilei confirms that his father came late to the study of mathematics, but with a twist. In this version, the roles are reversed. His grandfather *wanted* his father to study mathematics, but his father refused, until, finally relenting, he became interested, and then at last devoted to the subject. Perhaps Galileo had once told the boy Vincenzo this story as he tried to get him interested in mathematics, creating a fictitious example that his reluctant son never actually followed but at least remembered.

A close linkage between mathematics and the Renaissance arts, the very subject of this book, is stated explicitly by Viviani. That can only mean that this was the view of Galileo himself, and that he urged this view on Viviani. Viviani had won his place in Galileo's household as a mathematics prodigy, sponsored by the grand duke, but Galileo seems to have felt responsible for broadening Viviani's education beyond mathematics, pushing him to think creatively for himself and not to be satisfied with mathematics alone. When Galileo described his own education in the arts, linking it then to mathematics, he must have intended to pass along to Viviani something about how these things work together.

Interestingly, it is Gherardini who talks more about mathematics than Viviani, as if Galileo had also tried to broaden Gherardini's horizons, aware that he knew nothing about mathematics but confident that he would be interested in a philosophical way. Gherardini has this to say about Archimedes, on the basis of his conversations with Galileo:

> It is impossible to recount what benefit he received from studying this great man; it is certain that with his guidance he laid the firmest foundations and had no doubt that he would then lift himself on high, feathering the wings of speculation, investigating not just the most hidden operations of nature in this sublunar world below, but also the marvelous things of the upper celestial world. And one could, he said, travel securely without hindrance through heaven and earth, if one only did not lose sight of the teachings of Archimedes.[25]

These three little biographies do not overlap very much in what they choose to say, but they are unanimous in portraying Galileo's devotion to both mathematics and the arts. We can infer that Galileo frequently spoke about these things with all kinds of people, and that it was one of the most consistent and noticeable things about him. From the way he describes his education it is clear that in his mind mathematics and the arts were closely associated. As we investigate these things in the Renaissance context of what Galileo knew, we shall see how natural, and even inevitable, such an association is.

Geometry and the Arts

In his writing Galileo frequently juxtaposes talk about mathematics with talk about the arts, just as his first biographers describe. The association of mathematics and the arts is evidently not just something Galileo felt intuitively in his youth and recalled in his old age. It is to be found in his publications and his letters throughout his life. There is an extended example in the great *Dialogue* that was the subject of his trial before the Inquisition. In that ill-fated masterpiece Galileo immortalizes two friends, the Venetian Giovanni Francesco Sagredo and the Florentine Filippo Salviati, as well as an Aristotelian philosopher known here only as Simplicio, and not by his real name out of delicacy of feeling, as Galileo tells us in his foreword, "To the Discerning Reader." The dialogue takes the form of an extended conversation over four days. At the end of the First Day, Salviati, in a passage that was later cited by the inquisitors,[26] makes a remarkable claim about the nature of mathematical knowledge, "in which the Divine intellect indeed knows infinitely more propositions, since it knows all. But with regard to those few which the human intellect does understand, I believe that its knowledge equals the Divine in objective certainty, for here it succeeds in understanding necessity, beyond which there can be no greater sureness."[27]

"This speech strikes me as very bold and daring," gulps Simplicio. Not at all, says Salviati, and reminds his friends of some proofs in geometry that they surely know and appreciate, concluding that in such examples "I recognize and understand only too clearly that the human mind is a work of God's, and one of the most excellent."[28]

Sagredo bursts in with a heartfelt speech that superficially changes the subject, but only makes sense if it is somehow a continuation of the point about mathematics:

> I myself have many times considered in the same vein what you are now saying, and how great may be the acuteness of the human mind. And when I run over the many and marvellous inventions men have discovered in the arts as in letters, and then reflect upon my own knowledge, I count myself little better than miserable. I am so far from being able to promise myself, not indeed the finding out of anything new, but even the learn-

ing of what has already been discovered, that I feel stupid and confused, and am goaded by despair. If I look at some excellent statue, I say within my heart: "When will you be able to remove the excess from a block of marble and reveal so lovely a figure hidden therein? When will you know how to mix different colors and spread them over a canvas or a wall and represent all visible objects by their means, like a Michelangelo, a Raphael, or a Titian?" Looking at what men have found out about arranging the musical intervals and forming precepts and rules in order to control them for the wonderful delight of the ear, when shall I be able to cease my amazement? What shall I say of so many and such diverse instruments? With what admiration the reading of excellent poets fills anyone who attentively studies the invention and interpretation of concepts! And what shall I say of architecture? What of the art of navigation?[29]

In this passage mathematics and the arts seem to be connected, one calling up the other. Perhaps the first thing to do is to be sure that this is a real association of two things, mathematics and the arts, and not some spurious occurrence of two things together that do not really belong together.

As if to answer that suspicion, Galileo repeats this association near the end of the Third Day in almost the same terms that he used at the end of the First Day. He has been discussing *De Magnete*, a book by William Gilbert, physician to Queen Elizabeth I of England, describing simple experiments and observations on magnets, as well as the daring speculation that the Earth itself might be a giant magnet. Galileo's enthusiasm for Gilbert's book is nearly unbounded and quite unusual for him, as he usually takes very little notice of his scientific contemporaries. But despite nearly unlimited praise for Gilbert, the speaker in the dialogue, Salviati, says, "What I might have wished for in Gilbert would be a little more of the mathematician, and especially a thorough grounding in geometry, a discipline which would have rendered him less rash about accepting as rigorous proofs those reasons which he puts forward as true causes." To soften this criticism, which is perhaps unwarranted, given the newness of the subject, Galileo then spins out a metaphor that invokes the invention of the arts:

> I do not doubt that in the course of time this new science will be improved with still further observations, and even more by true and conclusive demonstrations. But this need not diminish the glory of the first observer. I do not have a lesser regard for the original inventor of the harp because of the certainty that his instrument was very crudely constructed and more crudely played; rather, I admire him much more than a hundred artists who in ensuing centuries have brought this profession to the highest perfection. And it seems to me most reasonable for the ancients to have counted among the gods those first inventors of the fine arts, since we see that the ordinary human mind has so little curiosity and cares so little for rare and gentle things . . . To apply oneself to great inventions starting from the smallest beginnings, and to judge that wonderful arts lie hidden behind trivial and childish things is not for ordinary minds; these are concepts and ideas for superhuman souls.[30]

Galileo's easy movement from mathematics, to invention in science, to the arts tells us that in his mind these matters are linked. In this passage "superhuman," applied to the hypothetical inventors of the fine arts, provides another bit of evidence. There is only one soul to whom Galileo routinely gave the adjective *superhuman:* Archimedes.

This metaphor of the arts is entirely characteristic of Galileo. Here he comments on the distinction between a philosopher who merely knows works of philosophy and a philosopher who knows Nature directly:

The difference between philosophizing and studying philosophy is that which exists between drawing from nature and copying pictures. In order to become accustomed to handling the pen or crayon in good style, it is right to begin by redrawing good pictures created by excellent artists. Likewise in order to stimulate the mind and guide it toward good philosophy, it is useful to observe the things that have already been investigated by others in their philosophizing; especially those which are true and certain, these being chiefly mathematical. But men who go on forever copying pictures and never get around to drawing from nature can never become perfect artists.[31]

Sagredo's lament is echoed, that so far from discovering anything new, he doubts that he can even copy what others have already done. The metaphor is the same as in the *Dialogue,* with mathematics and the arts mixed together inextricably.

The Many Meanings of Geometry

Geometry in the Renaissance meant many different things, and what Galileo meant by it changed over time, because by the time he published *Two New Sciences* he was using it to mean something it had never meant before. To begin to get at this complex cultural inheritance, I will survey in this section some of the meanings of geometry, and where in the culture these meanings were located.

To begin with, is it plausible that a well-educated, ambitious, intellectually curious young man like Galileo could be so unaware of Euclidean geometry that a belated encounter with it would change his life? When one considers that Galileo's father was a good mathematician who had even described the so-called Euclidean algorithm in print, it seems downright impossible. But confirmation comes from an incident in the least brief of John Aubrey's *Brief Lives,* the life of Galileo's near-contemporary, the English philosopher Thomas Hobbes:

> He was forty years old before he looked on geometry; which happened accidentally. Being in a gentleman's library Euclid's *Elements* lay open, and 'twas the 47 El. *libri* I [i.e., the Pythagoras theorem]. He read the proposition. "By"—he would now and then swear by way of emphasis—"G—!" said he, "this is impossible!" So he reads the demonstration of it, which referred him back to such a proposition; which proposition he read. *Et sic deinceps* that at last he was demonstratively convinced of that truth. This made him in love with geometry.[32]

Apparently it was possible to be well educated and still not know what Euclidean geometry was. It is noteworthy that Hobbes's encounter with it was accidental, but, as in Galileo's case, the love affair was life-long, taking the form in Hobbes's case of a fascination with certain notoriously difficult problems in geometry like squaring the circle, which he thought he had solved.

Why would a gentleman choose to display the Pythagoras theorem? The book must have been open like that of a dictionary or some other reference work, but Euclid is not a reference work, it is a textbook. It can hardly be the case that someone was looking up this proof and then left the book open. Rather it must have been on permanent display, like an intellectual decoration in the room.

Galileo's first encounter with Euclid was also accidental; in one version of it, he overheard Ricci lecturing to the court pages. In this case, too, Euclid was fulfilling a decorative function, as an intellectual ornament bestowed on the extended Medici household. What intrigued Galileo was, at least in part, what intrigued Hobbes, namely the singular certainty of the propositions of geometry, proved by irrefutable arguments. Such propositions were true in a peculiar and impressive way. In Galileo's case, though, geometry had a deeper significance than a mere collection of true propositions.

The legendary motto over the door of Plato's Academy, "Let no one enter without geometry," suggested that geometry was necessary for philosophy generally, but without any indication of what its philosophical role should be, or why it was essential. One thing was clear in the record: mathematics was a part of philosophy. Euclid and Archimedes were routinely called philosophers. Geometry occupied an erudite but apparently somewhat obscure corner of the Renaissance intellectual universe. It had a high status. It was something you might encounter among *court* pages or in a *gentleman's* library, where it had a static, ornamental quality. No one was proving new theorems in geometry. Geometry was a classical artifact with a vaguely sensed and uncertain philosophical importance. That was one of its meanings, but hardly a satisfying one. It was more of a question than an answer. This mystery might have been part of its fascination for Galileo. What lay behind Plato's door?

Geometry had re-entered European culture by way of translations from Arabic, beginning in the twelfth century. The Arabs in turn had made translations from Greek beginning in the ninth century. They also wrote their own geometry texts. The preface to a ninth-century Arabic geometry text captures the excitement, but also the elusiveness, of what this project was about.

Because we have seen that there are some things, a knowledge of which is necessary for this field of learning but which—as it appears to us—no one up to our time understands, and there are some things we have pursued because certain of the ancients who lived in the past had sought understanding of them and yet knowledge has not come down to us, nor does any one of those we have examined understand, and there are some things which some of the early savants understood and wrote about in their books but knowledge of which, although coming down to us, is not common in our time—for all these reasons it has seemed to us that we ought to compose a book in which we demonstrate the necessary part of this knowledge that has become evident to us.[33]

The recovery of classical geometry was pursued systematically by the Arabs and then by the Europeans, but there was a persistent vagueness about what this knowledge was for. The Greek texts, although written with model clarity, provided no hint of what one might actually do with geometry, even if one should achieve the knowledge and understanding that the sons of Mūsā, quoted above, were seeking.

The meaning of classical mathematics in the Renaissance was, for centuries, up for grabs. The mathematics of the Greeks had become available to Renaissance Europeans by way of translations, but what did this mathematics mean? The Renaissance had inherited an advanced body of knowledge without the user's manual, so to speak. It might as well have fallen from the sky. It was not clear what this sophisticated subject was for, nor what it could do. Europe was a nonmathematical culture in the possession of sophisticated mathematics.

Ancient mathematics had not been lost—on the contrary, enough of it survived to become, ultimately, the nucleus of a new science. Rather something else had been lost, the idea that you should *do* mathematics and *use* it, not just learn it as part of a closed, logical, philosophical system. Renaissance mathematicians, on the whole, did not do mathematics. Rather they acted as custodians for a finite body of material, a carefully circumscribed corner of philosophy. Both Galileo and Hobbes had studied philosophy, but they only encountered geometry elsewhere, by accident. Mathematics had survived as arcane, formal lore, but no one knew what the Greeks had originally been doing with it, and what the real questions were that had motivated their work. In fact, we still don't know that. Euclid's books tell us everything we need to know about his geometry except *why.*

Euclid's *Elements,* and all finished mathematics, can be understood as a purely logical construction, divorced from concrete meaning. It is possible to imagine that geometry developed in ancient Greece as a pure abstraction. In practice, though, that is not how mathematics works. Mathematics and all the sciences develop in response to good problems, to intellectual challenges that generate new concepts, methods, and ideas. There can be no question that Greek mathematics developed in response to concrete problems, and not merely abstractly, as later civilizations came to believe. The Renaissance itself is a proof of this. Inheriting classical mathematics as an abstraction, without its problems, without even the conception of a mathematics that could evolve in response to problems, Renaissance geometers invented no new mathematics at all.

The strength of mathematics is also its weakness—in its abstraction, mathematics says nothing about what it means, or whether it even has a meaning. What Galileo overheard from Ostilio Ricci was just such an abstraction. It must have seemed plausible in the Renaissance that mathematics had some deep and important meaning because, after all, the Greeks had created it with so much labor and care. But having studied it, each individual was as qualified as the next to form some private view of what that meaning was. It is not at all unusual to find in Renaissance writings a sort of "personal geometry" representing a private relationship to mathematics, a deeply held and idiosyncratic enthusiasm. Hobbes is an example, and we will see others.

On the other hand, it was also possible to believe that mathematics was genuinely unimportant and had no deep meaning at all. One of Galileo's opponents, Ludovico delle Colombe, wrote a small book *Against the Motion of the Earth* in which he ridicules the importance of mathematics, saying, "In Aristotle's day mathematics was considered a study for schoolboys, the first thing they learned, like our abacus school." Galileo has written in the margin of his copy, "The greater disgrace for this philosopher, who does not even know what every schoolboy knew in Aristotle's day."[34]

When Colombe talked about mathematics as the subject that little boys learn, he meant the arithmetic and practical geometry that was the standard education for boys of the merchant class in Renaissance Italy, quite distinct from the geometry of Euclid. This was geometry in a completely different sense from the philosophical one. Such mathematics had a clear use in business, surveying, construction, and similar practical applications. The tradition of such useful mathematics was essentially unbroken, dating back to classical times, and its meaning was not in doubt, but just for this reason it was of no philosophical interest. Colombe sneeringly suggested that Galileo's mathematics was nothing but abacus school mathematics. It is no surprise to learn that Colombe, from what little we know, was a frustrated social climber, jealous of his own status as a philosopher, in the face of a rival who, partly on the basis of his mathematics, had been named court philosopher to the grand duke of Tuscany. What Colombe meant by geometry was far from the geometry that had captivated Galileo, but it was a legitimate and familiar Renaissance meaning.

Galileo was professor of mathematics and taught geometry at the University of Padua until 1610, but the meaning of geometry in the university was something else again, a third meaning. An ancient tradition held that there were just four scientific subjects in all, the so-called quadrivium: geometry, arithmetic, astronomy, and music. Within the quadrivium these subjects were paired: geometry was the mathematics of astronomy and arithmetic was the mathematics of music. Although the university curriculum was no longer strictly organized in this way in Galileo's time, the tradition was not forgotten. A professor of mathematics, contrary to what one might expect, taught mainly theoretical astronomy. He would teach the geometry that was necessary for understanding a simplified version of Ptolemy's Earth-centered system of the world, and especially what was necessary to cast horoscopes. The clientele for such a course would be medical students. This is geometry that Galileo could have learned at the university, but it is geometry as an aid to astrological calculation, very different from the geometry of Ostilio Ricci.

While he was at Padua, Galileo taught this subject, but gave far more attention and energy to tutoring his private students. These lessons were distinct from his university lectures, and they involved yet a fourth kind of geometry. He wrote up notes for these students, some of which survive, and he even published one set of notes in 1606, his first book under his own name. *On the Geometric and Military Compass* was dedicated to the young Prince Cosimo de Medici, whom he had tutored the previous summer. The subject was a bit like abacus school geometry in that it was geometry applied to real problems, but Galileo's private teaching was aimed at young noblemen aspiring to military careers, so the tone was elevated. The military compass was a device of Galileo's own invention that could be used in several seemingly unrelated ways, for surveying a battlefield, aiming a cannon, or solving problems involving proportions and scaling, this last by a clever use of similar triangles. In his foreword to the reader, Galileo rather disingenuously alluded to a story about Archimedes and the young king of Syracuse, who wished to learn geometry, but couldn't spare the time it would take to travel that long road.[35] Was there not, he asked Archimedes, a royal road, not for commoners, but for him, that would be shorter and easier? Archimedes said, sadly, no, but Galileo said, well, now there is: my compass! Galileo deliberately conflates two very different notions of geometry. The Archimedes story was about Euclidean geometry, but the practical geometry that Galileo was offering to teach was a very different thing, a method of calculation, a sort of analog computing device. This was not the geometry that had so excited him at the age of nineteen, but he was willing to blur the difference in describing his private lessons. The classical reference was just too good to pass up, the perfect advertisement for attracting highborn students from all over Europe who would someday deploy their troops by geometry. Both Viviani and Gherardini believed that Galileo had tutored the Swedish king Gustavus Adolphus, one of the most feared commanders of the Thirty Years' War, in his supposed student sojourn in Italy.

Galileo was unusual in knowing all these different notions of geometry, and knowing them very well. But more typically and more often the different notions of geometry were mutually exclusive. The various uses of geometry did not really overlap, and someone who learned it from one point of view would have no interest in another one. Philosophers, even ones who knew Euclid, hadn't gone to abacus schools to learn the practical geometry of commerce. In the end, in *Two New Sciences*, Galileo created yet another meaning for geometry, a fifth meaning if we are still counting. The parabola, a curve of Greek geometry, became a metaphor for something physical, the arc of a projectile; and more generally, all geometry became, potentially at least, a metaphor for nature, promising that nature could be understood as geometry could be understood. It was not astronomy, and hence it was a new science, unexpected and startling. It employed geometry, but it had philosophical ramifications that seemed to explain the motto over Plato's door, as if for the first time.

Mathematics and Philosophy

The role of mathematics in philosophy is a controversy going back to Plato and Aristotle, and it makes a natural framework for considering Galileo's life and work (although as we shall see, another dichotomy is even more pertinent). If Plato had insisted on the importance of geometry, Aristotle very much downplayed it, to the point that his later commentators could argue that what looked like mathematics in Aristotle was only there in the service of other, more important ideas. Aristotle's philosophy strives to use the world's own nonmathematical terms to describe the world, and has at its best the appearance of brilliantly observed common sense. In Galileo's day the Aristotelian antipathy for mathematics was complete, as we have already seen in the case of Ludovico delle Colombe. Among the broader intelligentsia, though, like the Venetian salon society that Galileo frequented when he was professor at Padua (represented in his *Dialogue* by Sagredo) there was an audience and even an appetite for other views, making Galileo a favorite in this rarefied company.

However brilliant a figure Galileo might have cut among his friends in Venice, his status at the university, as professor of mathematics, was very low, while professors of Aristotelian philosophy were at the top of the heap. It is easy to infer how galling this must have been from Galileo's first book, published under a pseudonym in 1604, when he was already forty years old. A satire in Paduan peasant dialect, it lampoons the philosophy professors.[36] The two peasants in this little dialogue are trying to figure out the notion of parallax. They run about the woods and even climb trees, all the while making scurrilous remarks. They quickly figure out the basic phenomenon of parallax, and their only difficulty is in understanding why the philosophy professors don't get it (a pox on those goat turds at Padua!), and in particular why they don't understand that the "new star" of 1604 must have been much farther away than the Moon, since it showed no parallax.[37] It is not hard to hear Galileo's own frustrations in the scatological remarks of the peasants. Galileo put up with his low university position until, beginning in the summer of 1609 with the help of a certain Dutch invention that made distant things look nearer, he began to see his way out.

When Galileo succeeded in writing his own ticket at the Florentine court in 1610, he made sure that he was not just court mathematician but also court philosopher. This new, incomparably higher status gave him a platform that he had not had before. It also made him a target of Aristotelian philosophers at Florence and Pisa, as well as some philosophically trained members of the clergy, a problem that he had never had as a mere mathematician. Everyone remembers the Copernican controversy, in which the Church, after surprisingly little deliberation, declared the motion of the Earth not only contrary to Scripture, but also "foolish and absurd in philosophy."[38] This last was a reference not to Scripture but to Aristotle, as if Aristotle were also sacred. But there was another hard-fought controversy between Galileo and the Aristotelians in those years on the question of why things float, never really resolved in favor of either side. (It was at a debate on floating that Galileo first won the admiration of the visiting Maffeo Barberini, who would later become Pope Urban VIII, remembered today chiefly for his anger against Galileo in 1633.)

Although these controversies with the Aristotelians were notorious at the time and have been a staple of the Galileo story ever since, Galileo never spoke other than respectfully of Aristotle himself. He faulted the Aristotelians of his day, but argued against them that in light of new evidence their teacher would have changed his mind. "If they [ancient authorities—meaning Aristotle] had seen what we see, they would have judged as we judge," was the way he put it in the *Sunspot Letters* of 1613, where the topic was whether the Sun was inalterable.[39] In a 1640 letter to Fortunio Liceti, professor of philosophy at Padua, he wrote,

I claim (and surely believe) that I observe more religiously the Peripatetic, or I should rather say Aristotelian, teachings than do many who put me down as averse from good Peripatetic philosophy; and since one of the teachings given to us admirably by Aristotle in his *Dialectics* is that of reasoning well, arguing well, and deducing necessary conclusions from the premises, when I then see conclusions deduced that have no connection with the premises, which therefore deviate from the Aristotelian doctrine, I think I may rightfully deem myself a better Peripatetic.[40]

Gherardini reports a summary opinion of Galileo on ancient philosophy that is very instructive. It corroborates what Galileo says in the above letter to Liceti without the rhetorical cleverness, but it also says a good deal more.

It was not at all true, he said, that he had spoken with scorn and disdain of the ancient philosophers, namely of Aristotle, as some of those who profess themselves his followers had falsely claimed. He said only that the philosophical methods of this great man had not succeeded, and that there were errors and fallacies in them. He praised certain works, like the books of the *Hypermenia* [*sic*] and above all others he praised the *Rhetoric* and the *Ethics*, saying that in these arts he had written wonderfully. He praised Plato to the skies, for his truly golden eloquence, and for his method of writing and composing in dialogue form; but above every other he praised Pythagoras for his method of philosophizing, while he said that in his genius Archimedes had surpassed them all, and he declared him his teacher.[41]

Galileo's attitude toward Aristotle is consistent in his publications, letters, and informal conversation, and it is not quite the adversarial stance that one might have expected. Meanwhile, what he says about Plato might seem quite odd. He praises Plato for his literary style and for his dialogues (as his father had also done). But there is not a word for Plato's faith in mathematics.

Although Plato was fascinated with mathematics and prescribed it, for example, to the philosopher-kings of *The Republic*, it is not clear that he knew much mathematics himself. The Platonism of Galileo's day, better called Neoplatonism, was an elaboration of this attitude, but without much mathematical content, a peculiar idealism and number mysticism for which Galileo had no sympathy at all. In calling attention to Plato's literary style and pointedly not mentioning what was most distinctive in the Neoplatonism of his day, he was signaling, as clearly as anyone could, his distaste for that particular mathematical strain in Renaissance culture.

Pythagoras, meanwhile, is praised "for his method of philosophizing," a bit surprising, since in the popular Renaissance view Pythagoreanism, or Neo-Pythagoreanism, is essentially a synonym for Neoplatonism. That cannot have been what Galileo meant. Galileo seems to associate Pythagoras (but not Plato) with the method of mathematics in philosophy, because the mention of Pythagoras then calls up Archimedes. In Renaissance culture, so intensely interested in the classical past, figures of antiquity efficiently represented a kind of shorthand for present attitudes, both at a superficial level and in more studied ways. Galileo had to explain in his letter to Liceti what he (and only he) meant by Aristotle, for example. To investigate what Galileo meant by Pythagoras, we must make the first of many excursions into the classical past, with an eye to how the classical legacy manifested itself as it was imported into the world of the Renaissance.

2 The Classical Legacy

What did Galileo mean by praising Pythagoras "for his method of philosophizing"? At the very least Galileo must have had his own, private conception of Pythagoras, just as he had his own, private conceptions of Plato and Aristotle. Since Galileo's "method of philosophizing" was his life's work, Pythagoras is cited here in a way that demands attention.

The Pythagoras of tradition founded a utopian community in Croton (in southern Italy), where he and his followers lived a life governed by philosophy.[1] The very words "philosophy" and "mathematician" are supposedly coinages of the Pythagoreans, the *mathematikoi* being the inner circle of that exclusive society. The *mathematikoi* actually did mathematics at a very high level, as we shall see. The *akousmatikoi* (auditors) were the outer circle. All Pythagoreans were subject to strict rules of dress, diet, and behavior. They are said to have been vegetarians, a tradition at odds with another tradition, which says that Pythagoras sacrificed one hundred bulls to the gods in thanks for his famous theorem. The distance between these two stories should give some indication how vague our knowledge of Pythagoras and Pythagoreanism is.

Plato visited Pythagoreans in Sicily, long after the death of Pythagoras, on as many as three different journeys, evidence for his interest in

Pythagoreanism. The emphasis on mathematics in Platonic philosophy is just one example of Pythagorean influence on Platonism. Plato's *Republic* itself seems to describe something like a Pythagorean community. Plato's friend Archytas of Tarentum was a living, breathing philosopher-king, even if that was not his title.[2] Plato's enthusiasm for Pythagoreanism, together with the decline and eventual disappearance of the Pythagorean communities, came to mean that what survived of Pythagoreanism was Platonism. They had merged. Yet, as we have seen, Galileo distinguishes them. That is what is so odd in Gherardini's report.

One of the earliest sources to discuss Pythagoreanism critically is Aristotle, who wrote a book, now lost, called *Against the Pythagoreans.* In the *Metaphysics,* he mentions the Pythagoreans in several places, mainly to refute them. He dislikes their emphasis on numbers, but what he knows of it is trivial. Alexander of Aphrodisias, one of Aristotle's early commentators, quotes from Aristotle's lost book on the Pythagoreans: "they called the number five 'wedding,' since a wedding is a coming together of male and female, but in their view male is odd, while female is even; but this is the first number that originates from two (as the first even number) and three (as the first odd number)."[3]

If critics of Pythagoreanism trivialized its mathematics in such ways, Neo-Pythagoreans by the Roman period were doing much the same thing. They apparently agreed that "Number is the first principle, a thing which is undefined, incomprehensible, having in itself all numbers which could reach infinity in amount. And the first principle of numbers is in substance the first Monad, which is a male monad, begetting as a father all other numbers."[4]

Even some modern scholars seem to accept that Pythagorean mathematics consisted of such stuff, basing their view on surviving texts that are explicitly called Pythagorean, texts that are mainly about the Pythagorean way of life, its ethics, and its rules.

It is possible to imagine that what is being presented as Pythagoreanism in *all* the surviving materials is merely the view of the *akousmatikoi*, the outsiders. There were some notable leaks from the inner circle, the *mathematikoi*, genuinely interesting mathematics as we will see below, but the Neo-Pythagoreans seem to have no knowledge or awareness of real mathematics. They are like *akousmatikoi*, following the rules of the sect, perhaps, but not participating in the philosophy. Some of the later Neo-Pythagorean texts seem to be rationalizations for rules whose intent was no longer understood, as if the inner circle had disappeared and the outer circle was trying to make up for its absence.

The original Pythagoreans were notoriously secretive and exclusive. Admission to the inner circle required a rigorous five-year initiation ordeal. Anger on the part of some who were not admitted supposedly led to the sect's destruction in Croton and its dispersal, so we can be pretty sure that standards were not lightly compromised. One must consider the possibility that Pythagorean secrecy was essentially maintained throughout the sect's entire history. The evidence for this is not to be found in any Pythagorean text, of course, by definition, but rather in the mathematics that survived the end of the old Pythagoreanism, clear evidence for something remarkable going on behind the closed door.

Archytas, for example, a known Pythagorean and one of the *mathematikoi*, was credited by the late classical Greek commentator Eutocius with an astounding construction that is analogous to the compass and straightedge constructions of plane geometry, except that it takes place in three dimensions, and constructs curves in space as intersections of surfaces. With these curves he solves the problem reportedly posed by the oracle at Delos of duplicating the cube, that is, constructing the side of a cube that has twice the volume of a given cube. To put this kind of mathematics on a par with Monads and Dyads is to miss a crucial distinction. Yet when Aristotle, and even Plato, talked about Pythagorean mathematics, they could not make this distinction, presumably because they never gained access to the secrets of the inner circle. They were at best *akousmatikoi*. They only knew what anyone might know, because the *mathematikoi* guarded the doorway. It seems clear that even Plato never gained entrance. His own legendary doorway, supposedly requiring geometry to enter, is perhaps a substitute for the one that he wasn't allowed to enter himself.

Galileo is entirely aware of the distinction between *mathematikoi* and *akousmatikoi*. In his great *Dialogue*, right at the beginning, he addresses this point. Salviati has just said, to the surprise of Simplicio, that he does not see anything particularly wonderful about the number three.

> *Simplicio:* You, who are a mathematician, and who believe many Pythagorean philosophical opinions, now seem to scorn their mysteries.
>
> *Salviati:* That the Pythagoreans held the science of number in high esteem, and that Plato himself admired the human understanding and believed it to partake of divinity simply because it understood the nature of numbers, I know very well; nor am I far from being of the same opinion. But that these mysteries which caused Pythagoras and his sect to have such veneration for the science of numbers are the follies that abound in the sayings and writings of the vulgar, I do not believe at all. Rather I know that, in order to prevent the things they admired from being exposed to the slander and scorn of the common people, the Pythagoreans condemned as sacrilegious the publication of the most hidden properties of numbers or of the incommensurable and irrational quantities which they investigated . . . Therefore I believe that some one of them, just to satisfy the common sort and free himself from their inquisitiveness, gave it out that the mysteries of numbers were those trifles which later spread.[5]

Salviati is Galileo's spokesman and alter ego, and he here confirms that Galileo did indeed believe "many Pythagorean philosophical opinions." That he ridicules Neo-Pythagoreanism does not at all contradict this idea, but only makes it precise. One wonders if Galileo knew a fragment of Archytas that expresses his own self-reliant view: "To become knowledgeable about things one does not know one must either learn from others or find out for oneself. Now learning derives from someone else and is foreign, whereas finding out is of and by oneself. Finding out without seeking is difficult and rare, but with seeking it is manageable and easy, though someone who does not know how to seek cannot find."[6]

The Pythagoreans were secretive, but a few of their greatest mathematical discoveries leaked out, being perhaps just too exciting to keep under wraps. Besides the Pythagorean theorem, which was never secret and may not even have been Pythagorean, two of these discoveries exerted a particular fascination in antiquity and in the Renaissance. Both are mentioned by Plato, a very early source, proof that they are genuinely Pythagorean discoveries and that they became public at an early date.

Incommensurability

In a story that has many variations, the Pythagorean Hippasus was supposedly drowned for revealing one of the deepest Pythagorean secrets, the existence of incommensurable magnitudes. Incommensurability didn't find a completely satisfying resolution until modern times, in the construction that we now call the *real numbers*. In this language, what Hippasus revealed was the existence of *irrational numbers*, a discovery alluded to by Salviati, above.

The discovery was made by considering a simple figure showing the diagonal of a square. The startling fact is that there is no "ruler" that could measure both the side and the diagonal of the square exactly. That is, there is no smaller length that would go p times into the side and q times into the diagonal, where p and q are integers. This is what it means to say that the side and the diagonal are incommensurable.

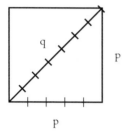

No unit evenly measures both the side and the diagonal of a square.

In the figure it looks as if there might be a unit such that $p = 5$ and $q = 7$, but by the Pythagorean theorem this would require $q^2 = p^2 + p^2$, and clearly 49 is not $25 + 25 = 50$. The Pythagoreans showed that there is *no* way to do it. For if there were, we would have integers p and q such that $q^2 = 2p^2$. We may assume that p and q have no common integer factor, because if they did, we could divide each of them by that factor and eliminate it without changing the relationship. Now q^2 is evidently an even number, because it has 2 as a factor, and that means q itself is even. Thus q^2 contains 2 as a factor at least twice, and hence p^2 contains 2 as a factor at least once; i.e., p^2 is even. But then p itself is even, and q is also, contradicting our initial remark that p and q would not have a common factor. The assumed relationship has led to an absurdity, and hence there cannot be such p and q.

The fact of incommensurability means that there are proportions that don't correspond to what we usually mean by fractions. It isn't enough to consider just proportions like 2 : 1, 3 : 2, and so on. There are irrational proportions, like the diagonal to the side of the square, that we do not have a simple way to express with numbers. This mysterious difficulty acquired almost religious overtones in the Renaissance, but it also had practical significance in the theory of music and in Galileo's mathematical physics, as we shall see.

The Platonic Solids

Another Pythagorean (or perhaps Hippasus again) got into trouble for revealing the existence of the *dodecahedron*, a solid figure with twelve regular pentagons for its faces.

The dodecahedron is perhaps the most surprising of the five regular polyhedra. These are the convex solid figures having congruent faces that are all the same regular polygon, arranged the same way at each vertex. That there are exactly five of them is a Pythagorean discovery. The most familiar of these regular polyhedra is the *cube*, having the square for a face, with three squares meeting perpendicularly at each vertex (corner), for a total of six faces in all. If we use an equilateral triangle for the face, we can make three faces meet at each vertex, forming the *tetrahedron*, with four faces in all; we can make four faces meet at each vertex, forming the *octahedron*, with eight faces in all; or we can make five faces meet at each vertex, forming the *icosahedron*, with twenty faces in all. These four, together with the only regular polyhedron having a pentagonal face, the dodecahedron, form the complete list of the five regular polyhedra, now called the Platonic solids because of what Plato suggested about them.

In the dialogue *Timaeus*, one of the founding documents of Neoplatonism, Plato makes a wild attempt, surprisingly influential in the Renaissance, to suggest what it could mean for Nature to be essentially mathematical. He imagines that the four elements, as they were then understood—Fire, Water, Earth, and Air—might have as "atoms" regular polyhedra, giving them their characteristic properties. The atoms of Water would be icosahedra, the most nearly round of the polyhedra, able to roll on each other as water does. The atoms of Earth would be cubes, suitable to make solid structures. The atoms of Air would be octahedra, and the atoms of Fire would be tetrahedra, with their sharp points. Nearly two thousand years later Galileo still found it a useful picture to think of fire as consisting of sharp tetrahedral particles, able to cut apart the atoms of solids, that is, to melt them. *Timaeus* speculates that in natural processes the polyhedra might dissociate into their constituent triangles and recombine, in this way making new compounds, since most things contain not just a pure element, but varying proportions of all four.

An attentive reader might object that there are five Platonic solids, not four. That is true, of course, but we have not yet mentioned the material of Heaven, which is different from the four elements. Heavenly material has for its atom the fifth solid, the dodecahedron. This remarkable speculation confers enormous mystical importance on the dodecahedron, with consequences that we will meet later. And for all its speculativeness, proposing a mathematical atomic theory in the fourth century B.C.E. still looks more than a little inspired.

Some very high-level mathematics from around the time of Plato did not survive. Of Archytas and Eudoxus, we know that they were superb mathematicians, but their work has been lost, surviving, if at all, in the words of others. It is only from the founding of Alexandria and the golden age of Hellenistic mathematics that we have a coherent corpus of material, including a systematic treatment of the earlier Pythagorean results.

The Hellenistic Age

The most convincing argument for thinking that the Pythagoreans were far more sophisticated mathematically than any account of them suggests is what happened in Alexandria, Egypt, as soon as it was founded, around 330 B.C.E. The Pythagorean communities had apparently disappeared by then, but excellent mathematicians immediately appeared in Alexandria, coming not from Athens or any known cultural center but from some other part of the Greek-speaking world to initiate a golden age of Greek mathematics and mathematical exposition, beginning with the works of Euclid. The sudden emergence of Alexandria as a world center of learning is quite remarkable. From an early date Alexandria was the center of Greek mathematical science. Throughout the Hellenistic Age, Alexandria served as a kind of hub, with its great Museum and Library the most important research center. Scholars like Archimedes were educated in Alexandria and continued to correspond with the center even when they settled elsewhere. The Hellenistic mathematicians did not call themselves Pythagoreans, but it seems inescapable that they were the heirs of the renowned Pythagorean mathematical tradition. They are neither Platonic nor Aristotelian, seeming to be both more mathematical and less religious than either of these other philosophies.

These Hellenistic mathematicians don't talk about philosophy at all, a truly unusual feature in their work, even if we just compare it with other Greek mathematics. Perhaps, though, their mathematics *was* their philosophy. That would be consistent with what we know of the old Pythagoreanism. When Galileo invokes Pythagoras, he seems to mean not just Pythagoras himself, of whom he knew as little as we do, but also his Alexandrian descendants. In Gherardini's recollection, Pythagoras and Archimedes are mentioned together.

Archimedes, Apollonius, and Euclid, the three Hellenistic mathematicians cited by Galileo in *Two New Sciences*, were near contemporaries, all flourishing in the Hellenistic Age, bracketed roughly by the founding of Alexandria around 330 B.C.E. and the death of Archimedes in 212 B.C.E. In this short period, hardly more than a century, Greek science reached astonishing heights of sophistication. We know this not from secondhand reports; in fact, there is a peculiar absence of second-hand reports. No one tells us about this culture until its spectacular end under Cleopatra VII in 30 B.C.E. Rather, we know the scientific sophistication of Alexandria from its primary scientific materials that survived and gradually became comprehensible again.

Our own history tells us that mathematical science develops with explosive rapidity when mathematics is coupled to practical problems, commerce, invention, and experiment, and, correspondingly, practical arts develop very rapidly when they can use mathematics to solve problems by pure thought. It has been suggested that the only plausible explanation for the rapid achievement of the Greeks is that something similar took place in the Hellenistic Age.[7] Their mathematics must have developed at least in part in response to practical questions, even while, in its intellectual sophistication, it generated its own new theoretical questions.

The myth told and retold later, by the uncomprehending civilizations that inherited their materials, is that the Greeks were simply in love with abstraction and didn't care about practical applications of their mathematics. This is not very plausible, and it fails to explain why political developments—the conquest of Greek cities by Rome—brought abstract mathematics to a halt. It seems far more likely that abstract mathematics was just one facet of a lively creative culture that was cut short by the new Roman culture, less hospitable to innovation, a culture of standardization.

Euclid

We know virtually nothing about Euclid's life,[8] but he lived in Alexandria sometime around 300 B.C.E., soon after the founding of the city and a generation or so before Archimedes. Euclid's works seem to be textbooks for this new intellectual city, a compendium of known mathematical arguments and results put into systematic, logical form. That this mathematical culture was Pythagorean in its origins is clear from the contents of Euclid's greatest work, the *Elements*, and not just the famous theorem at the end of Book I. The final book of the *Elements*, Book XIII, describes the five regular polyhedra, a Pythagorean discovery. Book X implicitly assumes familiarity with the puzzle of incommensurability and contains various attempts to make sense of it that are still somewhat mysterious. Some of the number theory in Books VII and VIII is thought to be due to Archytas.

Book V contains material associated with the pre-Hellenistic mathematician Eudoxus, a transitional figure between the old Pythagoreanism and the Alexandrians. The topic is his sophisticated theory of proportions, an ingenious response to the puzzle of incommensurability. It is a measure of its sophistication that the key insight is found not in a theorem, but in a definition, Definition 5, capturing the subtle notion of what it means for two proportions to be equal. Another indication of its subtlety is that Definition 5 was not successfully transmitted in the translations to Arabic and then to Latin. The most commonly used Latin translation of Euclid in the early Renaissance, that of Campanus of Novara, had something quite incomprehensible where Definition 5 should have been,[9] and it was only sixteenth-century humanist scholarship working with more faithful Greek manuscripts that restored it. In his very last days, Galileo was writing a dialogue to explore Euclid's Book V, Definition 5, as we shall note again in a more fully developed context. The Eudoxian theory was not fully appreciated in modern times until the nineteenth century, when it became, in effect, the theory of the real numbers.[10]

The structure of Euclid's *Elements*, beginning with clear definitions and axioms, continuing with theorems and proofs, has been the model for good scientific and mathematical writing ever since. Archimedes used this style in his research articles, and Newton used it, loosely speaking, in his *Principia*. Galileo used it to some extent in *Two New Sci-*

ences, even though that book is ostensibly in dialogue form. It remains, in outline at least, the form of mathematical exposition to this day.

Euclid wrote books on optics, although all but the first has been lost. The sole surviving book, simply called the *Optics,* describes the geometry of naked-eye vision. It was often cited in Renaissance proofs of the correctness of the perspective construction in painting. It also addresses some unexpected questions, such as why things seem indistinct when they are very far away. A pseudo-Euclidean book called *Catoptrics,* meaning the optics of mirrors, survives. It uses the methods of the *Optics* together with the equal-angles law of reflection to explain what you see in a mirror, including curved mirrors. One of its theorems, for example, is that images seen in convex spherical mirrors look smaller if the sphere is smaller. The slightly casual statement is suggestive of a lack of rigor in this book. Its final proposition asserts that you can light a fire using a concave mirror to reflect the Sun.[11] The geometrical proof is far from convincing.

Archimedes

In 212 B.C.E., after eight months' siege, the little city of Syracuse in Sicily, which gave us not only the geometry of Archimedes but also the pastoral poetry of Theocritus, was overrun by a Roman army. The story is told in Plutarch's *Lives,*[12] and in the histories of Polybius and Livy. These accounts all agree on an almost superhuman role for Archimedes, whose war machines kept the Romans at bay until they finally gained entrance to the city by guile. According to Plutarch, the Roman general Marcellus was very interested in taking Archimedes alive, but the soldier who found him lost in thought, contemplating a geometry problem, killed him. His last words, as remembered in Latin, were supposedly *"Noli turbare circulos meos"*—don't disturb my circles. This story of how Archimedes had almost singlehandedly defended the city was very well known in the Renaissance, and Renaissance mathematicians could assume that their noble patrons knew the value of keeping a few geometers employed in case of military necessity, even if none of them knew very clearly what Archimedes had actually done. His story added considerably to the cachet of mathematics in the Renaissance.

Archimedes's inventions are spectacular. Both Plutarch and Polybius describe machines like cranes that could reach over the city walls (on the harbor side) and grab the attacking ships, lift them up by the bows, then drop them so that they sank or split. The Romans were so terrified of these devices that they would retreat at the first sight of anything raised above the walls.

What would we know of Archimedes's mathematics if we had only this story to go on? Cranes are essentially applications of Archimedes's law of the lever, which says that weights balance if the fulcrum divides the suspension beam in inverse proportion to the weights. We might correctly guess that Archimedes had a deep understanding of the principle of leverage, even if we didn't have his proof of the law. This elegant proof fortunately survives in a little treatise called "On the Equilibrium of Plane Bodies." In this case the story and the mathematics are consistent, and even reinforce each other.

Other parts of the story are not as easy to place, though. The defenders of Syracuse were supposedly able to send missiles flying in great numbers accurately over a variety of ranges. This hints at clever mechanisms for quick reloading, and possibly at a knowledge of the parabolic trajectory. There is no Archimedean treatise on motion, though, as we already know from Galileo, nor is there any secondary evidence that one ever existed. In view of the importance of the parabolic trajectory in situations like the siege of Syracuse, and in view of the simplicity of the law, we wonder again at the absence of all discussion of this problem. Might work on projectile motion have been a military secret? Perhaps not everything was written down in a form that would survive a military defeat. The Pythagorean tradition of secrecy might have survived in part even as innocuous topics in mathematics were freely discussed.

A much later version of the Archimedes story, appearing first, apparently, in Galen about 160 C.E., says that he designed mirrors to focus the sun's rays on the Roman fleet and set fire to the ships.[13] This is a wonderful addition to the legend, but it was certainly impossible then (it wouldn't be easy now). Galen wouldn't have suggested it if the phenomenon weren't known, however, and we know from Euclid's *Catoptrics* that burning mirrors were discussed scientifically at an early date. Burning lenses were certainly known as early as the fifth century B.C.E. Such a lens is mentioned in Aristophanes's comedy *The Clouds*, where Strepsiades tells Socrates that he could get a lawsuit dismissed by buying one of those "beautiful transparent stones at the pharmacy that they use to light fires." He would use it surreptitiously to melt the writing on the court clerk's docket.

Archimedes also wrote a treatise called *Catoptrica*, attested by at least two ancient writers,[14] but it is lost. According to Apuleius it discussed, among many other things, the use of concave mirrors to light a fire, and since it is Archimedes, it must have improved on pseudo-Euclid's proof.

Archimedes was clearly a master teacher, because so many stories survive about him, unique among the Hellenistic mathematicians, the way stories are fondly remembered of favorite professors: Archimedes at the siege, Archimedes in the bath, Archimedes singlehandedly launching a ship that all the townsmen of Agrigentum could not move. A story for which there is documentary evidence says that Archimedes sent his newest propositions to Alexandria but delayed sending the proofs, to give his colleagues there the pleasure of devising proofs for themselves. Now there is a teacher! And going even further, he would include one false proposition, to discourage anyone from claiming he had proved something when he had not. This is like an implementation of Archytas's injunction to find things out for yourself. Two thousand years later Galileo still responded to Archimedes's pedagogic genius by calling himself Archimedes's student, as he truly was.

Apollonius

The surviving books of Apollonius, *Conic Sections*, open with a chatty letter to someone named Eudemus. With a little updating of the mathematical topics, this letter could have been written by a scholar at a research institute today. Here is an excerpt:

> I don't believe you have forgotten hearing from me how I worked out the plan for these conics at the request of Naucrates, the geometer, at the time he was with us in Alexandria lecturing, and how on arranging them in eight books we immediately communicated them in great haste because of his near departure, not revising them but putting down whatever came to us with the intention of a final going over. And so finding now the occasion of correcting them, one book after another, we publish them. And since it happened that some others among those frequenting us got acquainted with the first and second books before the revision, don't be surprised if you come upon them in a different form.[15]

Everything in the letter suggests that Apollonius takes a pure mathematical pleasure in his new results, and the lively mathematical scene comes through with astonishing freshness. The books themselves are the most difficult in all of Hellenistic geometry, and, like Euclid's *Elements*, they do not discuss any applied problems. Of the eight books Apollonius describes to Eudemus, only seven survive. Among the other lost books of Apollonius is one *On the Burning Glass*, which it is conjectured might have dealt with parabolic mirrors.

When Galileo needed to invoke certain properties of the parabola in *Two New Sciences,* he didn't assume that his reader would know them. His character Salviati, in the guise of helping Simplicio, just rederives a few results of Apollonius on the spot.

Greek Applied Mathematics

In 1900 off the Greek island of Antikythera, a sponge diver discovered an ancient shipwreck at a depth of about 200 feet. Archaeological investigation the next year brought up amphorae that allow the wreck to be dated to the first century B.C.E., and also, rather spectacularly, a number of bronze and marble statues. The ship was apparently taking the booty of war back to Rome as Rome completed its conquest of the eastern Mediterranean. The most remarkable find was not immediately recognized, though. Only back at the National Museum of Archaeology in Athens were a few lumps of corroded bronze, encrusted with mineral deposition, recognized as the pieces of a kind of clockwork mechanism. It contained at least twenty interconnected gear wheels (more than thirty according to the most recent study, by X-ray tomography)[16] and several dials, with inscriptions for reading their positions. There were also inscriptions on plates which had originally been fastened to the box containing the device. Intensive work to understand what this mechanism was for proves that it is an analog computer for calculating the positions of the Moon and planets. It is by far the most complicated machine to survive from classical times.[17]

Nothing in the written record suggests the existence of any such device.[18] Its mechanical sophistication is completely unexpected. If it was a trophy of war, its value must have been clear not just to the Greeks who made it but also to the Romans who selected it to accompany the statues. Yet no one talks about anything remotely like it. Hero of Alexandria, a Romanized Greek writing perhaps 200 years later, our only real source for mechanisms in the ancient world, describes simple toys by comparison. The absence of texts on applied mathematics might suggest that Greek mathematics was purely abstract. The Antikythera mechanism suggests a very different picture. There was Greek engineering and applied mathematics, despite the absence of records.

It is virtually certain that Hellenistic mathematics was applied to real problems, but we have no documentation of how it was applied. What survives, Euclid's *Elements*, the treatises of Archimedes, and Apollonius's *Conic Sections*, are very abstract works at a very high level, and not in any condition to be put into immediate practice. The Romans knew of books (now lost) like that of Alexander's engineer Diades that described war machines. But if there were books that described how to apply the theory to make the machines, the Romans never saw them. As the Plutarch story tells us, the Romans were looking for immediately useful works at the sack of Syracuse and did not find them. Plutarch, a Romanized Greek, ends the story of Archimedes by lamenting that Archimedes had written only theoretical books and had left no practical instructions for building his machines. Perhaps this is exactly the point, though. Perhaps the Greeks guarded the technological advantage they had, even as they lost ground across the eastern Mediterranean, by keeping this information out of Roman hands. Plutarch, together with Roman civilization in general, seems unaware that the real secret was in the theoretical works. They talk as if Archimedes's methods had died with him.

Plutarch had actually read and understood at least some of Archimedes's works. We can be quite sure of this, because no one who had not understood them could describe Archimedes's works so accurately: "For it is not possible to find in geometry more profound and difficult questions treated in simpler and purer terms. Some attribute this success to his natural endowments; others think it due to excessive labour that everything he did seemed to have been performed without labour and with ease. For no one could by his own efforts discover the proof, and yet as soon as he learns it from him, he thinks he might have discovered it himself; so smooth and rapid is the path by which he leads one to the desired conclusion."[19] Yet in spite of his understanding of Archimedes's achievement, Plutarch still does not make the connection to Archimedes's machines. It is as if the geometry and the engineering had nothing to do with each other, in his mind. The Romans seem not to have the concept of applied mathematics. As far as they are concerned, mathematics is just theory.

Mathematics changed under Roman administration, in the direction of becoming divorced from practice. Whether by accident or by design, Hellenistic mathematics, in all its abstraction, was becoming, by about 100 C.E., a kind of code, concealing its great practical potential. Apparently not realizing its importance, the Romans never even tried to decipher it. Understood this way, the job of the Renaissance mathematicians who inherited the Greek treatises was to crack the Greek code, to translate theory into practice. The recurring suspicion that powerful secrets must be hidden there was literally true.

In hindsight it is clear that the most useful scientific discovery of antiquity was Hellenistic geometry. What later civilizations actually inherited, though, was a confusing mix of scientific subjects and styles. We see, also in hindsight, that the height of sophistication reached in the Hellenistic Age was not maintained. Rather than advance, scientific texts began to degenerate back toward earlier, simpler ideas.

In this process of degeneration, Hellenistic mathematics became subordinated to Aristotelian philosophy, assigned a restricted place in the accepted view of the universe. How that happened in detail is unknown. The sources that might tell us, documents from Cleopatra's Egypt, for example, don't survive. The result was an unstable legacy, with Aristotelian orthodoxy, later transmuted into Christian dogma, wrapped around and containing Hellenistic mathematics. This "philosophy" couldn't last forever. There was always the possibility that Hellenistic mathematics would break through and resume its independent role, as, ultimately, in the scientific revolution of the seventeenth century, it did.

Plutarch's On the Face in the Moon

A tantalizing vignette from the period of degeneration survives in a fragment of dialogue by Plutarch from about 100 c.e., *On the Face in the Moon.* The discussants in this enigmatic work consider why the Moon has the features that we see, and more generally what the Moon is. They stroll and talk, ornamenting their ideas with apt quotations from the poets. They discuss mythic qualities of the Moon, and whether there might be inhabitants there. And although this is not, by their own account, a particularly scientific group of speakers, they also cite scientific ideas otherwise unknown to us from antiquity. For a modern reader the wide-ranging freedom of the scientific discussion takes on an elegiac quality, one that it wouldn't have had at the time, in our knowledge that we are witnessing something that was about to disappear.

Lamprias, who is believed on the basis of other Plutarch dialogues to be Plutarch's brother, leads the discussion, but everyone is keen to see that all points of view be represented, including ones that are quickly refuted, such as that the Moon only seems to have markings because it dazzles and confuses our eyes by its brightness. Lamprias and Lucian, a philosophical colleague of his, defend the idea that the Moon is much like the Earth, and that the markings represent high and low places, light and shadow, against others who suggest that the Moon is like a mirror and that what we see in the Moon is the Earth reflected. These arguments would recur centuries later in the aftermath of Galileo's telescopic observations of the Moon, and once again it would be debated how a rough surface scatters light, in contrast to how a mirror reflects light. In Plutarch's *Moon,* Lucian suggests that on a rough surface a ray of light can reflect multiple times in the interstices of the material, so that it emerges at an angle different from what the simple equal-angles law of reflection would predict, a nice argument. He also suggests, however, that light reflected from a concave mirror violates the equal-angles law, a very different conception from the pseudo-Euclid treatment of concave mirrors, which applies the equal-angles law at each point of the mirror.

The Stoic participant, Pharnakes, argues that the Moon must be made of some light material like air or fire, since it doesn't fall down to Earth. Oddly, the Aristotelian idea that the Moon is made of a special celestial material, something that naturally stays up and moves in circles, is not represented. Lucian, in arguing that a heavy, Earthlike Moon would not fall down, makes an amazing analogy to a stone whirled in a sling, which also does not fall, "for each thing is governed by its natural motion unless it be diverted by something else,"[20] almost a paraphrase of Newton's laws of motion. On the other hand Lamprias, who takes Lucian's part, seems to ridicule the Aristotelian idea of a spherical Earth, or at least one that is formed by an attraction toward the center. The idea that humans might be standing upside down on the other side of the Earth, or that we ourselves walk slanted like drunkards, seems laughable to him, as if the direction *down* were absolute in space and not just toward the center of the Earth. Thought experiments are described (to point out how absurd they are) of masses dropped down through the center of the Earth, shooting through the center, then turning around and coming back, attracted again toward the center, evidence that someone else must have taken such notions seriously. All in all there are hints of a scientific culture much richer than what actually comes down to us.

The astronomers Hipparchus and Aristarchus are mentioned, the latter in connection with his idea that the Earth moves, one of very few surviving bits of evidence for this idea in antiquity. Fifteen hundred years later Copernicus, in the dedication of his revolutionary book to

Pope Paul III, would cite Plutarch on this point, and even copy some of Plutarch's Greek into his own book, as if to assure the reader that it was really true that a few Greeks, at least, had thought that the Earth moved.

How close this idea came to disappearing completely is evidenced by Copernicus's choice of an urbane Greek essayist as his authority.

Ptolemy's Almagest

The Arabs called it *Almagest*, "The Greatest," Claudius Ptolemy's second-century tome describing the universe as a whole. It is detailed and mathematical, and uses voluminous tables of observational data, including Babylonian eclipse data, together with a geometrical model of circles on circles, to give a fairly accurate description of the motions of the planets and the stars. Ptolemy famously describes these motions as if from a stationary Earth, and even begins with arguments about why the Earth *must* be stationary.

Almagest contains elements of originality, but on the whole it is not an original work. Ptolemy promises "we shall only report what was rigorously proved by the ancients, perfecting as far as we can what was not fully proved or not proved as well as possible."[21] That is, *Almagest* merely updates, with possible improvements, an earlier work in the same vein, the Hellenistic astronomy of Hipparchus, which is occasionally quoted in *Almagest,* but which otherwise has been lost.

Most curiously, Ptolemy is very deferential to the philosopher Aristotle in his opening remarks. The more one thinks about it, the less sense this makes. Even Aristotle mentioned his opponents' points of view, if only to refute them, but in Ptolemy we must infer the opposing points of view, because he does not tell us about them. All those reasons why the Earth must be stationary, for example, tell us that there must have been those who thought the Earth was *not* stationary.[22] Otherwise why bring it up? Ptolemy never actually points this out, and his failure to show us both sides of the argument is truly unfortunate. We are left to speculate endlessly what the real status of this question was in the ancient world, the question that would resurface with such acrimony in the Galileo affair.

More generally, in the invocation of Aristotle, Ptolemy adopts a moralizing tone, a religiosity almost, that is simply not there in Euclid, Archimedes, or Apollonius. Here is how *Almagest* begins. "Those who have been true philosophers, Syrus, seem to me to have very wisely separated the theoretical part of philosophy from the practical. For even if it happens that the practical turns out to be theoretical prior to its being practical, nevertheless a great difference would be found in them; not only because some of the moral virtues can belong to the everyday ignorant man."[23] Ptolemy seems eager that his work not be considered practical. The tone is a chummy one, as between "true philosophers," and the implication is that we already agree on a question that actually drops upon us without warning. Among other things, a kind of parallelism seems to be assumed, with "practical" and "ignorant" on one side, and "theoretical" and "virtuous" on the other. Never mind that these things don't sound parallel to our ears. This introduction to an ostensibly mathematical treatise comes loaded with moralizing baggage.

Ptolemy goes on to say, now that he has safely established his work as theoretical, that "Aristotle quite properly divides the theoretical into three immediate types: the physical, the mathematical, and the theological."[24] Mathematics occupies a separate space between physics and theology, where it deals only with eternal things. To put it more plainly, the physical and the mathematical are mutually exclusive categories. A mathematical physics, a mathematical description of things on Earth, is impossible. There can be no such thing, because on Earth things decay and die, and are not eternal, as mathematics is.

The association of mathematics with Heaven in Ptolemy's version of Aristotle's philosophy and the separation from physics had truly peculiar consequences over the next millennium and more. Aristotelian philosophers simply didn't waste their time on mathematical theories of earthly things. On the other side, theology and mathematics were associated very closely. As a result, both Islam and Christianity acquired the same plan for the universe and the afterlife, and both of them essentially got it from Aristotle. Mathematics, eternity, and theology were already linked in a pre-Christian philosophy. The familiar idea of Heaven as a place "up there" is essentially to be found in Aristotle's *On the Heavens.* Ptolemy's *Almagest* develops this picture in mathematical detail, adopting Aristotle's view that the substance of Heaven is unlike anything on Earth, being perfect and unalterable.[25] The mathematical counterpart to this perfection is that all heavenly substances move in the perfect way, the circle. Christianity and Islam adopted this picture in its entirety.

Aristotle is an early source, pre-Hellenistic, yet here, in the Roman period, he is back on center stage, laying down the law in one of the most influential mathematical treatises of all time. It is impossible to know exactly how this reversion to earlier ideas occurred, but it seems clear that citizens of the Roman period found Aristotle congenial, with his piety, his static and systematic sorting of things into categories and types, and his encyclopedic coverage of all philosophical subjects. The absence of serious mathematics also made Aristotle easier to read, no doubt. The Romans did not intentionally do away with Hellenistic scientific work, but under their administration it became infused with moralizing, and its standards changed markedly. It became an enterprise subordinate to philosophy, and vigorous debates about competing pictures of the universe seem to have vanished.

It is especially strange to find Aristotle in a mathematical treatise. Aristotle is implicitly somewhat hostile to mathematics, but in *Almagest* even the very mathematical Ptolemy makes arguments that limit the role and the importance of mathematics. He asserts that mathematics does not apply to the highest part of the Cosmos (God, theological), nor to the lowest part of the Cosmos (Earth, physical, practical), but only just in the middle (the observable Heavens, virtuous, mathematical).

That typically three-part Aristotelian conception had a very stultifying effect on both mathematics and physics. Galileo still confronted it some fifteen hundred years later. This restriction on mathematics may have been untenable in the long run, but it actually did stop both mathematics and physics for a very long time, a circumstance of our history that is almost difficult to believe. The reverse transformation, the replacement of these Roman-era ideas with conceptions of mathematics closer to the Hellenistic ones, took place only in the Renaissance, and only very slowly.

Perhaps it is too harsh to blame the Romans, except in an incidental way, for developments within philosophy that remained Greek. Few Romans could have comprehended these questions, or could have cared. A tendency to moralize with regard to mathematics is not peculiar to Aristotelians. Plato too assigns to mathematics a peculiarly moral role. As we have noted, the education of the philosopher-kings in *The Republic* includes a crucial role for mathematics, as a step toward that true understanding that they will need as rulers. What is surprising in the Roman-era *Almagest*, written some 500 years after Plato, is that it seems to be assuming the attitudes of that earlier period, attitudes that in the intervening Hellenistic Age were nowhere to be seen.

Hellenistic mathematics, which is essentially Pythagorean, represents a simpler, less encumbered view of mathematics. Hellenistic mathematics says nothing at all about morality, or what the significance of mathematics might be. Yet in its careful logic and scrupulous clarity it embodies a steadiness that might well be taken for a virtue, and being Pythagorean it is implicitly ethical—just not in a way that is spelled out.

The Renaissance inherited these contradictory ideas of what mathematics was, and what it ultimately meant, without necessarily even appreciating how contradictory they were. The result was to make a vague and confusing inheritance even vaguer. The Aristotelian tradition assigned mathematics an inessential role, but still a very exalted one, in its limited function of describing the observable Heavens. The Platonic tradition gave mathematics almost unlimited importance in theory, but did not actually do real mathematics with any success. In Neoplatonic guise, it tried to find cosmic significance in trivialities. The original Pythagorean *mathematikoi* did amazingly high-level mathematics but made no particular claims for it, at least to outsiders, and in fact left very little on their own account, so that we have to depend on others to know anything about them, even when their reports are hostile or ignorant.

Music and Number Theory

One of the most charming stories of Pythagoras tells how he discovered the mathematical nature of music. He passed by a smith's shop and heard the hammers on the anvils producing musical sounds. Going in, he found that these sounds did not depend on the way the hammers were swung, but only on their respective weights. He went home and experimented with strings suspending different weights. When twelve pounds hung on one string and six pounds on another, and he plucked them and listened, they sounded an octave apart. This, he realized, was because the weights were in the ratio 2 : 1, and the ratio 2 : 1 became forever after associated with the octave. Similarly, when one string held twelve pounds and the other eight, that is, weights in the ratio 3 : 2, then he heard the interval which is called a musical fifth, and the ratio 3 : 2 became associated with the fifth. The ratio 4 : 3, eight pounds and six pounds, gave the musical fourth, and so on.

The story of Pythagoras and the smithy is told in the early fourth-century *Life of Pythagoras* by Iamblichus, nearly one thousand years after Pythagoras actually lived. We might expect it to have no credibility at all, but amazingly it reports a true scientific observation, although in garbled form. It is in fact true that two tones, one octave apart, are in the ratio 2 : 1 *in frequency*. Middle C, for example, is 256 cycles per second, and C an octave higher is twice that, 512 cycles per second. Simple experiments, repeated in the Renaissance in the attempt to rediscover the Pythagorean knowledge, produce this ratio in connection with the octave. But it is *not* true that *weights* in the ratio 2 : 1 stretching equal strings will produce an octave (weights would have to be in the ratio 4 : 1). It seems inescapable that such experiments had indeed been done by the Pythagoreans, who first discovered the wonderful connection between mathematics and music, but that by the Roman period it was a rather dim memory, and no one knew exactly how it worked. The details are stated wrongly, in just the way one would expect of someone who had heard the idea but didn't understand it, and hadn't checked. The idea itself, though, that the ratio of the weights controlled the relative pitches of the vibrating strings, is a sophisticated one. It is mathematically scientific at a high level, although the discovery that it relates is just on the edge of being prehistoric.

This observation about arithmetical ratios and musical intervals is interesting enough, in a scientific way, but in Plato's books it assumes a cosmic, even supernatural significance as "the music of the spheres." The universe is number, and number is music, and somehow these things must be understood all together. The idea is to be found, among other places, in Plato's *Republic*, in a vision of the universe in which the planets turn on the Spindle of Necessity: "Upon each of its circles stood a Siren, who was carried round with its movement, uttering a single sound on one note, so that all the eight made up the concords of a single scale."[26] The belief in a cosmic harmony, the music of the spheres, became the single most characteristic belief of Neoplatonism. It must have arisen out of Plato's enthusiasm for Pythagorean ideas and not from anything Pythagoras or the *mathematikoi* believed themselves, since there was never any outcry about yet another secret being revealed, but it proved remarkably durable. Galileo's great contemporary Johannes Kepler was not just a believer in the music of the spheres, he felt that this music had inspired his greatest discovery.

Ptolemy, who wrote the *Almagest*, also wrote a book of music theory, the *Harmonics*, that was very influential in the Renaissance, intensely studied by Galileo's father, Vincenzo Galilei, and by Kepler. Like the *Almagest*, Ptolemy's *Harmonics* opens with a long-winded philosophical preamble on the theoretical and the practical. Eventually it gives us a thorough and elaborate account of the musical scales based on corresponding arithmetical ratios, but it also relates these notes to the motions of the planets, an attempt, apparently, to put the music of the spheres on a scientific footing, and to understand more deeply, through its astrological influence, the effect of this music on the human soul.

Iamblichus's source for the story of Pythagoras and the smithy was a different treatise of the same period, the *Manual of Harmonics* by Nichomachus of Gerasa, a Romanized Neoplatonist and contemporary of Ptolemy. In chapter 3 of his *Manual* Nichomachus suggests, with Plato, that the notes of the musical scale are probably derived from the planets, "For they say that all swiftly whirling bodies necessarily produce sounds when something gives way to them and is very easily vibrated; and that these sounds differ from one another in magnitude and in region of voice either because of the weights of the bodies or their particular speeds." Nichomachus's mention of the planets' weights as possibly determining their notes sounds like an allusion to the story of Pythagoras and the weights. He promises to tell us in another book "the reason why we ourselves do not hear this cosmic symphony which emits a complete and all-harmonious sound, as tradition reports," but we will never know—that book does not survive.

Aristotle had considered the music of the spheres in *On the Heavens*, and had even politely suggested that it was a charming conception and not so implausible. After all, when large things move, on Earth at least, they typically do make sounds. But in the end, he dismisses the music of the spheres very sensibly. How do we know that the planets don't make sounds? Because we don't hear them.[27]

Arithmetic and Number Theory

We do have another book of Nichomachus, his *Introduction to Arithmetic*. This *Arithmetic* starts out, in language rather like Ptolemy's, with several chapters of observations on eternal matter, the meaning of life, the way to philosophize properly, and other things that do not seem to have much connection with arithmetic, before finally getting to its topic in chapter 7, odd and even numbers. We must make a conscious effort to realize that the philosophizing *is* by now part of the subject. Mathematics has become an innocuous corner of philosophy.

The book contains facts about numbers, but without any proofs, so that one doesn't actually learn mathematics from it, if by mathematics we mean a way of thinking. A few of the facts are sufficiently striking that one might remember them, such as that the sums of successive odd numbers are squares: $1 + 3 = 4$, $1 + 3 + 5 = 9$, $1 + 3 + 5 + 7 = 16$, and so on; and 4, 9, 16, and so forth are just the squares of 2, 3, 4, and all the rest. Understanding such intriguing relationships among the integers (whole numbers) is the subject of number theory, a lively part of mathematics even today. And this particular example even played a role in Galileo's discovery of the parabola law, although he wouldn't have learned it from here. In Nichomachus's hands, however, number theory is basically a tedious and disorganized compilation, without proof, of Pythagorean results from centuries earlier, aiming to make you a better person by putting you in touch with eternal verity.

A far better source for Pythagorean number theory in the Hellenistic sense is of course Euclid's *Elements*, Books V–X, which, although it is some 400 years earlier than Nichomachus, contains not only coherent arguments with proofs, but also much more sophisticated subject matter. When we compare Nichomachus with Euclid, we find in the Roman-era work the same phenomenon we have noted in the case of Ptolemy's astronomy, a peculiar infusion of moralizing philosophy, and in Nichomachus's case also a precipitous decline in mathematical sophistication.

A still earlier source for Pythagorean number theory is Aristotle himself, but as we have already noted, Aristotle was quite hostile to Pythagoreanism and clearly trivializes it. According to Aristotle, reporting at the level of the *akousmatikoi*, the Pythagoreans believed "number is the essence of all things," and "assumed . . . the whole heavens were harmony and number."[28] The "music of the spheres" seems to be a version of what Aristotle is reporting, possibly standard lore of the *akousmatikoi*.

The Latin Legacy

The works we have described to this point were in Greek, but in *Timaeus*, the Platonic dialogue that popularized the Platonic solids as atoms, we meet a Greek work that was actually translated into Latin, by Marcus Tullius Cicero. The vague picture of triangles coming together to form matter was somehow appealing, and this undemanding picture of the atoms probably became familiar to the small class of Roman literati. In another work, the *Dream of Scipio,* Cicero has left a vision, also very Platonic, of the universe as a whole, with the spheres of the planets circling the Earth, as seen from above. Scipio hears the music of the spheres up there, of course, and tells us that the ears of men "overcharged with this sound have grown deaf to it," reminding him of an African tribe living close to the roaring cataracts of the upper Nile who are, by this circumstance, completely deaf.[29]

Cicero also gives a charming description of finding the tomb of Archimedes at Syracuse, about 140 years after the siege, which he recognized by the figure carved into the stone, illustrating Archimedes's own favorite theorem: the volume of the sphere to the volume of the containing cylinder is in the ratio 2 : 3. Cicero reports that he had to tell the Syracusans what a great man lay buried there, as they did not even know. We may infer that mathematics in Syracuse had ended, although there was no reason to doubt this.

Two Roman authors, Vitruvius and Lucretius, write about Greek science in their own words, and even though their understanding is surprisingly primitive, their books are still interesting because they, like Plutarch, draw on texts that we don't know from other sources. One wishes that they had made actual translations. Their books, as part of the classical scientific heritage, were influential in the Renaissance, although in very different ways.

Vitruvius's *On Architecture* comprises ten books that touch on a great many things, even astronomy. Architecture draws on many arts and sciences, and Vitruvius is acquainted with many Greek results. In telling us about them, however, he occasionally betrays almost unbelievable ignorance. He knows about Eratosthenes's determination of the radius of the spherical Earth, for example, but it turns out that he thinks that the Earth is a disk, with ourselves in the middle of it, and that Eratosthenes has determined the distance from us to the edge of the disk.[30] Vitruvius's text was nonetheless popular in the Renaissance and much studied, and it is fun to read.

It is from *On Architecture* that we get the story of Archimedes running naked through the streets shouting "Eureka!" As Vitruvius tells the story, King Hieron had given gold to a goldsmith to make a crown, but when he received the finished crown, he suspected the goldsmith of cheating him. The weight of the crown was correct, but what if the smith had kept some of the gold for himself and made up the lost weight in less-precious silver? He asked his friend Archimedes to investigate. Archimedes, as he was contemplating this problem, lowered himself into a bath and saw how the water ran over the side, corresponding to his immersed volume. Eureka! If the crown had more volume than an equal weight of gold, then the crown could not be pure gold. He immersed the crown, collecting the displaced water, and then immersed the same weight in gold, proving the goldsmith's forgery.[31]

Another original Latin scientific text is from Lucretius, who wrote a long and beautiful philosophical poem called *On the Nature of Things*. Its basic premise, taken from Epicurean philosophy, is that the world is made of atoms that move about randomly and interact by chance. It is explicitly atheistic. As a contrast to the very religious philosophies that survived into the Christian period, it is interesting to see this irreligious one. Lucretius has nothing but contempt for people who find comfort in religion and cannot accept "the nature of things." Nothing guides the atoms to any purpose. This identification of atomism with atheism meant that such atomistic speculation was essentially off-limits in the Renaissance. (That did not prevent Galileo from publishing a few atomistic speculations anyway.)

It is worth noting that Lucretius, like Vitruvius, believes that the Earth is flat, and he wonders how the Sun gets back around underneath the Earth during the night, even to the point of suggesting that it is perhaps a different Sun that rises each morning.[32] These Romans, even though they are acquainted with Greek learning to some extent, seem to live in a different universe, quite literally.

Vitruvius and Lucretius are interested in Greek lore, but only in the answers, which they sometimes misunderstand, being unacquainted with the questions. In their work we glimpse the process by which Greek science became uncoupled from the practical problems that generated it, to appear to later generations as an almost magical abstraction.

The Roman Empire is said to have ended, if one must assign a date, in the year 476 with the surrender by Romulus to Odovacar the

Visigoth, who declined to become emperor. Theodoric the Ostrogoth was similarly uninterested in the title after personally assassinating Odovacar in 493. But now, with Rome well and truly finished, a small miracle occurred. At this unlikely moment, a few Greek mathematical books were at long last translated into Latin.

The remnants of cultured Roman society seem to have noticed, finally, that classical learning was about to be permanently lost. Their most brilliant representative, a young man of good family named Boethius, was educated in Greek and did what he could, at this very late date, to save something. He chose to translate Nichomachus's *Manual of Harmonics,* his *Arithmetic,* and a few propositions of Euclid's *Elements,* but unfortunately he did so in the style of Nichomachus, that is, without the proofs. He also translated several books of Aristotle. By itself this couldn't have done very much, and it didn't even go on for very long. But now another miracle occurred.

Boethius was highly regarded by Theodoric, because his learning helped Theodoric impress other Gothic kings with the sophistication of his court. Like many another favorite, though, Boethius was eventually accused of plotting against his sovereign and was executed, but in Boethius's case not before he had managed to compose a remarkable book, the *Consolation of Philosophy.* This book, literally the dying breath of classical Latin, seemed to demonstrate through the awful history of its composition the strength and worth of classical learning. It is beautifully and intelligently written, calmly argued, proving in its own balanced style the truth of what it is saying.

The *Consolation of Philosophy* begins autobiographically with Boethius himself, fallen from the heights of honor and esteem to the pitiable state of an unjustly condemned man, but it soon moves to examine the great philosophical questions of good and evil, justice, and the purpose of life. It is not overtly Christian, but it is not inconsistent with Christianity either. Rather it extols virtues that are admired in every culture: courage, balance, and an aspiration to ever-higher understanding. It was translated into Old English and Old High German, then Middle English (by Chaucer) and every other European language. It became perhaps the most popular book of the millennium. In his death Boethius had accomplished the impossible. He had fired future generations in the West to learn what was so powerful in the classical legacy.

Poetry

Dante's description of being in love, in the *New Life*, is poetically experimental, clinically analytical, and emotionally extreme. Somehow it manages to be all these things at once as it tells the story of his love for Beatrice, whom he first saw at the age of nine and whom he loved from that moment on. For Dante, we realize, as we grapple with this strange book, poetry is not just a pretty arrangement of words on a page but rather an intellectual tool of vast capability, a method for the deepest problems, an instrument that he has begun to master.

In the disasters of his life, not just the death of Beatrice at the age of twenty-four but also his own exile from his native city Florence, he turned to Boethius. And, as he says, "it sometimes happens that a man goes seeking silver, and, beyond his expectations, finds gold." Dante plunged himself into the study of classical learning, and soon began to write his own version of the consolations of philosophy, with the title *The Banquet*. Like Boethius, he is guided by Lady Philosophy, who has, for the moment, replaced Beatrice.

The Banquet was never finished: mere learning was not enough for Dante. Erudition that, for some, would be a suitable end was for Dante only the beginning, the raw material for his poetic imagination to create, out of his extraordinary knowledge, a new world, so powerfully rendered and so real that we recognize it even today, the world of *The*

Divine Comedy. From the inscription over the Gate of Hell, "Abandon all hope, you who enter," to the reunion with Beatrice above the Earthly Paradise, to the vision of God at the end of the poem, Dante sets himself impossible poetic challenges, then surmounts them in language of crushing finality in Hell and celestial sweetness in Heaven. Along the way we meet so rich a world of every description that a fair approximation to the question, what did anyone know in the year 1300? is, what did Dante know?

Readers are sometimes annoyed that the language in the last volume, *Paradiso,* becomes increasingly geometrical. The imagery of the circle is everywhere, and there are even theorems of Euclid. With a proper appreciation for the intellectual roots of the poem, though, we need not be surprised. The importance of mathematics for classical culture was, if anything, overrepresented in Boethius, and we have seen how deeply intertwined mathematics and philosophy had become, even while mathematics preserved an aura of secret power no longer comprehensible. Would Dante have ignored this mystery when he did not shrink from any other challenge? Hardly—this was a job for poetry! Geometry takes its sublime place at the end of the poem by the carefully structured choice of the poet.

The two passages I will consider in the next chapters are very startling indeed. They were not even recognized as mathematics until the twentieth century,[1] being beyond the mathematical experience of most readers even today, and, as one would expect, they are not in standard mathematical language. Unrecognized, they never functioned as contributions to mathematics *per se,* but they contributed in every generation to the penumbra of mysterious importance that surrounded mathematics in the Renaissance and nourished it, because *The Divine Comedy* was known and loved wherever Italian was read. Now that we can read these passages as they were intended, they give us unexpected insight into the richness of speculation that accompanied the classical legacy, and the high level of mathematics possible in Europe even before the Renaissance, hundreds of years before Galileo.

3   The Plan of Heaven

Dante's *Paradiso* is built on the plan of Aristotle's *On the Heavens*. That Christians might conceive of the world this way was established in the theology of Thomas Aquinas in Dante's own lifetime. Aquinas doesn't explicitly describe Aristotle's universe, but he adopts its terms in speaking about the arrangement of things, which amounts to an implicit acceptance of it. He considers the question of the Earthly Paradise, for example, and whether it might be located as high up as just under the sphere of the Moon. He concludes that it could not be that high, because the element Fire rises to that region, and would burn it, something the Venerable Bede, who originally made the suggestion, seems not to have realized.[1]

In Aristotle, and also implicitly in Aquinas, our world of the four elements extends only as far as the sphere of the Moon. Here begins the realm of the fifth element, in successive spheres that carry the heavenly bodies in order: Moon, Mercury, Venus, the Sun, Mars, Jupiter, and Saturn.[2] Above all these is the sphere of the fixed stars, the eighth sphere, and last comes the Primum Mobile, the ninth sphere, the one that turns all the others. These nested spheres surround the stationary Earth of Aristotle's cosmology. Above them is the Empyrean, the abode of God and the angels.

Just as Dante journeys through the circles of Hell in the *Inferno*, and climbs up the cornices of Mount Purgatory in *Purgatorio*, here in *Paradiso* he and Beatrice ascend sphere by sphere through the heavenly regions, stopping in each one to speak with the spirits of the blessed. The Aristotelian universe functions for Dante like a magnificent stage set. Each of the heavenly spheres has its own character, and is the appropriate home for some particular category of blessed spirit. The sphere of the Moon, for example, the lowest sphere, is the heaven of the Inconstant, those who, for some pardonable reason, broke their holy vows. It is fitting for the Inconstant to be here because the Moon is changeable in its phases, as centuries later Juliet would admonish Romeo,

> O, swear not by the Moon, th' inconstant moon,
> That monthly changes in her circled orb.

If it occurs to you that the Moon, despite being part of immutable Heaven, might still be just a little bit changeable because it is close to the Earth, where everything is subject to chance and change, then you are getting into the spirit of the Aristotelian universe, and have begun to feel its seductive attraction as a system.

The ascent through the Aristotelian universe works naturally to structure the poem until Dante and Beatrice reach the Primum Mobile, but now the poet has a problem. This is as far as the Aristotelian universe goes. It is not at all clear what the Empyrean, the region beyond, should look like. No one has ever described it before.

The Hypersphere

Dante could be pardoned for just ending the poem right there, with some kind of mystical vision that gives up on geometry, but that is not what he does. His Empyrean is both geometrical and precise, and totally surprising. At the end of Canto XXVII of *Paradiso*, he looks down at the Earth from the Primum Mobile, seeing it far below, surrounded by the planetary spheres of the Aristotelian universe. At the beginning of Canto XXVIII he turns his eyes the other way, upward, and sees, geometrically, much the same thing. A bright geometrical point, representing God, is surrounded by circular ranks of angels, nested around God just the way the planets are nested around the Earth. Dante and Beatrice are at the point *A* shown in the figure, in the middle on the Primum Mobile, where the two systems touch.

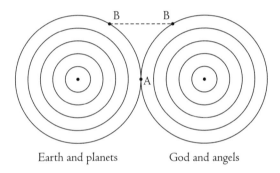

Earth and planets   God and angels

From the Primum Mobile one can look down toward the Earth or up toward God, whether one is at point *A*, or point *B*, or any other point.

What the figure cannot faithfully show is the symmetry of this universe. It looks in the figure as if there is only one way to ascend from the Earth to God, the way Dante and Beatrice came, through the point *A*, but in fact any ascent from the Earth is in the direction "up" and also leads to God. Dante is careful to say, in Canto XXVII, that there is no special point in the Primum Mobile, that the Primum Mobile is uniform everywhere. Thus what the illustration seems to show, that there is a special point *A* on the Primum Mobile halfway between the Earth and God, is a bit misleading. There is nothing special about the point *A*, shown in the middle where the two systems are *depicted* as touching. The two systems actually touch along the entire Primum Mobile, including at the point *B* for instance, which is a single point despite being shown in two places. Every point of the Primum Mobile is halfway between the Earth and God. Every point on the boundary of the lower universe coincides with a corresponding point on the boundary of the upper universe. The dotted line is our attempt to indicate that the two points labeled *B* are really the same point, but we choose not to distort the spheres in order to bring them together. Rather we simply remember that when we leave one system of spheres at *B*, we enter the other one at *B*. There is no "edge" to tell us that we have moved from one of the systems to the other, just as Dante encounters no edge when he continues upward into the Empyrean at *A*. This universe is therefore finite in volume, having twice the volume of one of the two spherical systems, and it has no edge—an idea that seems paradoxical until we study the figure and see how it could work.

Dante is clearly delighted with this conception and wants to make sure that we get it. He describes it in several different ways, or rather Beatrice does, explaining it to Dante. One detail that they discuss is the speed with which the various spheres are turning. In the background to this discussion is Aristotle's proof that the universe must be finite in size. As Aristotle argues, the Earth is stationary in the middle, and all the other spheres basically turn around the Earth once in twenty-four hours (diurnal motion), carrying the stars and planets through the sky. There are in addition all the particular motions of Ptolemy that give each planet its peculiar motion, but the diurnal motion is the main one. Now a small sphere, like the sphere of the Moon, doesn't have to move particularly fast to get around in twenty-four hours, but the larger the sphere, the bigger its circumference, and thus the farther the corresponding planet has to travel to get all the way around. This means that the larger spheres must move faster. If the universe were infinitely large, it would have to move infinitely fast, an absurdity. In

Canto XXVIII Dante, knowing that the spheres below him move faster the larger they are, is puzzled to see that the angelic spheres of the Empyrean turn faster the *smaller* they are (and thus the closer to God, the point at the center). Beatrice says, in effect, that their speed indicates their *height*, not their size, with higher spheres turning faster, just like the planetary spheres below. Thus the whole system shows a marvelous consistency (she says). And, in an ingenious way, Dante has incorporated Aristotle's requirement that the universe be finite, without giving it an edge.

To see this idea another way, it helps to consider a two-dimensional analogue that is more familiar, namely the spherical surface of a ball, like the surface of the Earth. Starting from the South Pole, at 90° south latitude, we can imagine the circles at 80°, 70°, and so on up to the Equator. These circles enclose the Pole, and grow in size as we go northward. The largest of these circles, the Equator, encloses the entire Southern Hemisphere. If we continue northward, into the Northern Hemisphere, the circles of constant latitude begin to grow smaller again, until we reach the northernmost point, the North Pole. In this process the latitude circles first grow, then shrink, with height, if by "height" we mean how far north they are.

As an application of this idea, one may map the Earth on flat paper by dividing it into southern and northern hemispheres and mapping these hemispheres separately, much as we do in the previous figure, except that now the two-dimensional diagram represents a two-dimensional surface, not a three-dimensional space. One sometimes does see maps of the Earth laid out like this. Where we say "Earth and planets" in the Plan of Heaven figure we could say "South Pole and Southern Hemisphere," and where we say "God and angels" we could say "North Pole and Northern Hemisphere." We understand that the largest circle in either hemisphere is the Equator, and that it is the common boundary of the two hemispheres. The two hemispheres are glued together along the Equator. Dante's universe is just like this, except that it is three-dimensional, not two-dimensional, and the two "hemispheres," which are really three-dimensional balls, are glued together along the Primum Mobile. This three-dimensional analogue of the two-dimensional spherical surface is called a *hypersphere,* or a 3-sphere, for short.

The branch of mathematics that deals with spaces like this, spaces that are different from the space that we visualize most easily, is called *topology.* In visualizing a new three-dimensional space, finite but having no edge, Dante has invented a new topological space, the 3-sphere. If this imaginative feat had been recognized in his own time, and if the idea had been pursued and developed, Dante would today be considered one of the inventors of topology, and one of the great creative mathematicians of all time. As it is, he is not even a footnote to topology, which was only invented officially in the eighteenth century, and didn't really take off until the twentieth.

Geometry in Paradiso

The 3-sphere in *Paradiso* is an interesting clue to the state of mathematics in the early Renaissance. It tells us, first of all, that Dante is a first-rate mathematician, and therefore a trustworthy witness on questions of mathematics, perhaps our best source. His understanding of mathematics, with allowance for his genius, must be the early Renaissance understanding.

The hardest thing for a modern person to understand about the 3-sphere, as a model of the universe, is that it is finite, that it doesn't go on forever. This contradicts the Euclidean model of space, the space of our present-day common sense, in which straight lines *do* go on forever. How was Dante able to get past this common-sense objection and even conceive of a finite space?

Dante doesn't necessarily consider Euclid's geometry, with its straight lines, as a model for space. Dante thinks of geometry as a branch of philosophy, consisting of propositions that are true with mathematical certainty, because their proofs are there to read. But these true propositions don't necessarily say anything about real space. Nowadays that seems like a very sophisticated conception, because it is only in the twentieth century that we have broken free from Euclidean geometry to consider non-Euclidean models of the universe (the 3-sphere was Einstein's favorite model). In Dante's case, however, it perhaps represents a lack of sophistication. Euclidean geometry had become detached from its roots in common sense, and it made very little claim to describe anything. Dante understands Euclid in this rather unspatial way. Dante cites two theorems of Euclid in *Paradiso*, and neither describes anything spatial. Both are cited merely as examples of things that are known to be true with certainty.

The first Euclidean theorem in *Paradiso* occurs in Canto XIII.[3] Dante and Beatrice are in the Sphere of the Sun (the Intellectuals). Fittingly, they hear Thomas Aquinas discoursing on wisdom, and especially on Solomon, who asked God for the wisdom to govern his people well. Aquinas points out that there were many kinds of wisdom that Solomon did *not* ask for, and among these he includes geometry. Solomon did not ask to know whether there could be any triangle inscribed on the diameter of a semicircle that was not a right triangle. (If Solomon had known his Euclid, he would not have had to wonder about this.) This casual, almost dismissive reference to Euclid suggests that Dante knows Euclidean geometry very well, but doesn't give it undue importance.

The other Euclidean theorem in *Paradiso* occurs in Canto XVII.[4] Dante and Beatrice are in the Sphere of Mars (the Warriors). Dante addresses Beatrice with a question about his own future. Beatrice, he says, can tell what *may be* from what *must be* as surely as the human mind can tell that a triangle cannot have two obtuse angles. This is a reference to the Euclidean theorem that the sum of the angles of a triangle is two right angles. Since one obtuse angle is already greater than one right angle, a triangle clearly couldn't have two of them. The point of the reference, though, is not anything spatial. Dante is not talking about kinds of triangles, he is talking about kinds of knowledge.

Strange as it seems, geometry for Dante is less spatial than it is for us. Our common-sense idea that real straight lines go on forever comes from taking Euclid as a model for real space, but this is not common sense for him. When we think that space must be infinite, we are really thinking of a space made of Euclidean straight lines. This is apparently not such a natural idea for Dante.

Euclid himself may have been uneasy about what straight lines do at very great distances. After all, what knowledge do we have about this? The behavior of things at great distances from ourselves is not a part of experience. Euclid builds his whole geometry from five postulates, statements that are taken to be true without proof, as a starting point. In formulating these, he isolates the behavior of lines at great distance in the fifth and last postulate, called the *parallel postulate*, as if postponing having to take a stand on this matter as long as possible. Then in the development leading up to the Pythagoras Theorem at the end of Book I, he delays using the parallel postulate as long as possible. There was even a tradition of attempting to build Euclid's geometry without using the parallel postulate at all, or to substitute something more plausible for the parallel postulate. All of this is to say that our "common sense" about straight lines is not really so common. It has been viewed with suspicion even by people who consciously adopted it.

We will see in the next chapter that for Dante the favored geometrical entity was not the straight line but the circle. If we imagine building a space out of circles, not straight lines, a finite space becomes much more natural, and the 3-sphere is arguably the most natural of all. Dante could invent a non-Euclidean space because he didn't necessarily accept Euclidean geometry as a model for space. And this was not because he rejected it, but because it simply didn't have that status to begin with.

As a kind of confirmation that he was on the right track with the 3-sphere, his universe is much more natural theologically than a Euclidean universe. The angels sang "Glory to God in the highest," but where is the highest? In a Euclidean universe there is no highest point, but in Dante's 3-sphere there is. Not only that, but there is also a lowest point, at the center of the Earth, where Satan is fixed in ice. Thus the entire universe is spread out between God and Satan in a geometrically and theologically satisfying picture.

The idea that Dante could have done this has sometimes nonetheless been greeted with skepticism, and it is only fair to point out how overwhelming the evidence is. For one thing, the 3-sphere in *Paradiso* was noticed independently by readers again and again once the topological idea became widely known.[5] That is, the description of it is so clear that the prepared mind recognizes it at once, even though it seems so out of place. If so many people have noticed it, to their immense surprise, then it is plausible that it is really there.

For another thing, Dante is truly insistent that we see it and understand it, foreshadowing it, then describing it in more than one way, insisting that it is new, and even telling us how he did it. The foreshadowing comes at the end of Canto XXII, where Beatrice prepares Dante for what is coming:

> Beatrice began: "Before long thou wilt raise
> Thine eyes and the Supreme Good thou wilt see;
> Hence thou must sharpen and make clear thy gaze,
> Before thou nearer to that Presence be,
> Cast thy look downward and consider there
> How vast a world I have set under thee."[6]

Dante looks down, and being in the eighth sphere sees seven spheres below him, surrounding the Earth ("this little threshing floor"). He notes the sizes and velocities of the heavenly spheres, the very properties that he will make use of later in the analogy that turns this space, the one that he has already passed through, into the space that he is about to enter. At the end of Canto XXVII, in the ascent to the Primum Mobile, Beatrice instructs him to look down again. In ascending Dante is careful to say that every part of the Primum Mobile is like every other, so that we don't make the mistake of imagining that the point A in the Plan of Heaven were somehow special. At the beginning of Canto XXVIII Dante first sees the semi-universe ahead, but since he is looking down, he sees it reflected in Beatrice's eyes. This is in itself a good description of the 3-sphere: it is made of two halves, one a reflection of the other. Perhaps this is the way Dante first thought of it. Just as before, Dante now draws attention to the sizes and velocities of the spheres, this time the angelic spheres. In trying to understand what he is seeing, Dante explicitly refers to the two halves as the model and the copy ("l'essemplo e l'essemplare"). And in teasing him about the difficulty he is having, Beatrice says,

> "There's naught to marvel at, if to untie
> This tangled knot thy fingers are unfit,
> So tight 'tis grown for lack of will to try."[7]

That sounds like the assertion that this is a new conception, new mathematics that no one has ever understood before.

Some commentators have reacted to these lines with discomfort, as if to recognize creative mathematics in Dante would be inappropriate. This anachronistic attitude can only obscure a close reading of Dante in these mathematical places. In connection with this question, critics have called attention to *Paradiso* XXII, 67, "for it [the highest sphere] is not in space and does not turn on poles."[8] The implication is that none of the geometrical language about these things that apparently are in space and turning on poles is to be understood as describing a real space after all. With this suggestion I can only agree, but Dante's geometrical construction is unaltered by this observation. After all, even Euclidean geometry is not particularly spatial for Dante. When Dante emphasizes that heaven is not physical, he means that it is mathematical. For Dante *physical* and *geometrical* are not synonymous but opposed.

Dante himself perhaps says it best, in *Paradiso* IV, 37–42, as Beatrice explains to him that the blessed spirits in the lowest sphere are not really where they appear to be, but are only represented there:

> "They're shown thee here, not that they here reside
> Allotted to this sphere; their heavenly mansion,
> Being less exalted, is thus signified.
> This way of speech best suits your apprehension,
> Which knows but to receive reports from sense
> And fit them for the intellect's attention."[9]

Similarly the 3-sphere represents, as if to our senses, an almost inexpressible unity, going far beyond a literal representation. It is forgivable to be surprised at Dante's mathematics because virtually no one expects mathematical sophistication in a poem of the late Middle Ages. Apparently, though, this prejudice is wrong. Creative mathematics of a high order was not just possible, it was actually being done almost as soon as translations of classical mathematics were available. This circumstance must affect our view of the subsequent history of mathematics and science.

Formally Dante's mathematics can be identified with a modern construction in topology, but the meaning then and now of this mathematical object has changed completely. Dante's mathematics has to do with the heavens, as Ptolemy and Aristotle had insisted. Dante shows us how mathematical genius could have developed within that very restrictive framework. It might seem that the 3-sphere could only be an isolated example, and that such a "theological mathematics" could never be repeated. That would be selling Dante short: he invented theological mathematics not just once but twice.

4 The Vision of God

Having described what was never described before, the structure of the Empyrean, the abode of God and the angels, Dante might have brought the *Divine Comedy* to a triumphant conclusion, but no. This was not enough. He continues upward into the angelic regions, apparently heading toward the point that is God. He is blinded by the light, only to find that his eyes are given new strength, so that even here, where he should be overwhelmed, he continues to see. He seems to know where he is going. The poem concludes with a vision of God that is completely surprising and requires explanation. It appears to be, and to some extent it must be, a mystery. But, as has been suggested only recently,[1] it is based on one of the deepest works of Hellenistic geometry, Archimedes's *On the Measure of the Circle.*

The Last Lines of Paradiso

There can be no question that geometry plays some crucial role in the final lines of the *Divine Comedy*. There are circles, a turning wheel, and a simile involving an actual geometer, trying to "measure the circle." This is such a strange and unexpected figure that it may even spoil the climactic experience of the ending for many readers. To salvage this situation, and to reveal what Dante is really doing, may partially excuse the unpoetic treatment of the last lines, below. I translate the final lines fairly literally. They follow a description of a vision of the Trinity as three circles, with Dante's attention fixed particularly on the circle representing the Son:

> Within itself, of its own coloration
> I saw painted our own human form:
> So that I gave it all my attention.
> Like the geometer, who exerts himself completely
> To measure the circle, and doesn't succeed,
> Thinking what principle he needs for it,
> Just so was I, at this new sight.
> I wanted to see how the human image
> Conforms itself to the circle, and finds its place there;
> But there were not the means for that,
> Except that my mind was struck
> By a flash of lightning, by which its will was accomplished.
> Here strength for the high imagining failed me,
> But already the love that moves the Sun and the other stars
> Turned my desire and my will
> Like a wheel that is turned evenly.

Most editions of *The Divine Comedy* will have hundreds of explanatory and interpretive notes. At the lines on the geometer there is sure to be a note explaining the ancient problem of "squaring the circle." This is the problem of constructing, using the methods of Euclid's *Elements*, a square with the same area as a given circle, believed from classical times onward to be a very difficult problem, and finally, in 1882, proved to be impossible. It follows from the 1882 result that it is impossible to construct any polygon having the area of a given circle, using only compass and straightedge. (Here a *polygon* means any closed figure with straight lines for sides.) The reason is that if you could construct the polygon, then you could also construct the square with the same area. Euclid has methods for going from a polygon to a square. Thus the curved arc of the circle somehow cannot be transformed into the straight sides of any equivalent polygon by a Euclidean construction, and vice versa. A mere proof of impossibility does not stop the occasional enthusiast from trying to do it anyway, but it is a settled matter: you can't square the circle.

Knowing this much about squaring the circle still does not fully elucidate what Dante is saying, though. Footnotes typically go on to explain that just as the geometer cannot solve this problem in geometry, Dante cannot comprehend how the human becomes divine. Thus Dante is expressing a kind of futility, the impossibility of understanding a truth that is too deep. This interpretation seems plausible, and certainly not unexpected, given the high mysteries that are in play, but something still feels wrong about it. It is simply out of character for Dante to admit defeat. He has come this far, after all. Why, at this climactic moment, within a few lines of the end of his great poem, would he give up, and represent himself as not up to the last challenge, staring at a problem that he cannot solve? He has solved all the other ones, on the way to getting here. It just doesn't sound like Dante.

What is more, it is not so clear that he cannot solve the problem. There is that lightning flash that seems to indicate that he *can* solve it after all. Dante doesn't actually know that squaring the circle is impossible, since that wasn't proved for another 500 years. Is he perhaps suggesting that he can do it? That would surely be an unfortunate reading.

The poem does not at all end on a note of defeat. Thus the lightning flash must represent something important—perhaps, as many notes suggest, a transcendent insight into the drama of salvation, which is the larger issue here, Dante's own salvation being one of the extended themes of the poem. Maybe the geometrical image has been dropped at this point? And yet the geometrical image does seem to continue, even after the lightning flash, in the image of the turning wheel, as if the flash of insight were also a geometrical insight. It is very puzzling. To solve this mystery we must find out what Dante actually knew about geometry, and more particularly what he knew about this problem.

Dante's Geometry

Dante had access to essentially all of the classical geometry that survived. By his time both Euclid and Archimedes had been translated, more than once in fact, from both Arabic and Greek. Thus we in the twenty-first century have no particular advantage over him in terms of knowing classical geometry. Anything that we know, he could also have known. We have already seen that he does know Euclidean geometry and makes effortless use of it, although in a philosophical context that seems not to have too much to do with geometry the way we think of it now.

In Dante's philosophical book *The Banquet* he describes all the so-called Liberal Arts, including geometry. Here is what he says about geometry:

Geometry moves between two things antithetical to it, namely the point and the circle—and I mean "circle" in the broad sense of anything round, whether a solid body or a surface; for, as Euclid says, the point is its beginning, and as he says, the circle is its most perfect figure, which must therefore be conceived as its end. Therefore Geometry moves between the point and the circle as between its beginning and end, and these two are antithetical to its certainty; for the point cannot be measured because of its indivisibility, and it is impossible to square the circle perfectly because of its arc, and so it cannot be measured exactly.[2]

This is a truly strange characterization of geometry, quite idiosyncratic. It quotes Euclid, or rather it misquotes Euclid, who never says "the point is its beginning," or "the circle is its most perfect figure." Euclid simply lays out postulates and proves theorems, in the cleanest and most economical manner. In its strangeness, we can be sure that this passage will tell us quite a lot about Dante himself.

Nearly the whole passage is about things "antithetical" to geometry, and not about geometry itself, which is characterized by "certainty." Now the point and the circle really *do* belong to geometry, but they are treated here as if they didn't, and they seem to be singled out as therefore particularly interesting. All of this is a bit odd, but it is also oddly consistent with what we already know about Dante's geometry. The propositions of geometry are true with certainty: that is just how Dante has used them in Cantos XIII and XVII. He tosses them off rather casually in those cases, as if certainty didn't really impress him all that much, and here he comes right out and says, more or less, that it is the *uncertain* parts of geometry, the parts antithetical to certainty, that interest him.

The uncertain parts of geometry are the point and the circle. These are exactly the constituents of his image in Canto XXVIII, the plan of the universe as a whole. We already know that the point and the circle interest him intensely, and here he confirms that. Thus *The Banquet* seems to be a reliable guide to Dante's very personal and peculiar interpretation of geometry. Finally, Dante states clearly in this passage that the circle cannot be measured because of its arc, which is just the problem of constructing a polygon with straight sides in place of the circle's arc, here asserted to be impossible. That brings us back, face to face, with the defeatism of the geometer staring at a problem that he can't solve, now with Dante's own authority that he can't solve it. This just can't be right. Defeatism and Dante don't mix.

In all of Dante's writing he never mentions the name Archimedes, but there is a surviving treatise of Archimedes called *On the Measure of the Circle*. It is only a few pages long, and thus easily copied. It was widely circulated in Dante's day, and even well before. We have seen how intensely Dante was interested in the measure of the circle—it is inconceivable that he would not have known the Archimedes treatise. It is time to have a look at it.

The Archimedes Proof

The first thing to say is that Archimedes does not write a treatise on the measure of the circle only to say that he cannot measure the circle after all. He *does* measure it. This is a crucial point. In fact, it is Archimedes who found the formula for the area: πR^2. He does not of course *square* the circle, meaning he does not *construct* a polygon with the area of the circle. He does represent his formula by a triangle with the same area as the circle, however, even though the triangle is not constructible. We see that the problem of measuring the circle is subtler than a simple statement of impossibility.

An important part of Archimedes's argument can be represented in a sequence of figures, and anyone working through the treatise would draw these figures. We start with the given circle and introduce two perpendicular diameters, as in the first figure, opposite. Next, as shown there, we connect the sides of the square thus determined.

The measure of the circle, that is, its area, is now crudely approximated in this figure by the area of the square. The argument then goes on to approximate the true area of the circle with increasing accuracy by polygons having more and more sides, doubling the number of sides at each step.

The area of a polygon is easy to compute, because it can always be dissected into triangles. In this case, we can dissect the polygon of N sides into N congruent triangular wedges, as in the second figure. Each

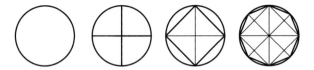

A sequence of figures from the Archimedes proof. The sequence continues infinitely, with each polygon after the square being replaced by a polygon with twice as many sides, just as the octagon at the far right replaces the square.

has an altitude h that is a little less than the radius R of the circle, and a base b that is the Nth part of the circumference of the polygon. The area of the wedge is then exactly $(1/2)hb$ and, taking this N times, the area of the polygon is exactly $(1/2)hbN$, or one-half the altitude h multiplied by the circumference Nb of the polygon. It would be tempting to say that as N goes to infinity, the altitude h of the polygon becomes R, the radius of the circle, and the circumference Nb of the polygon becomes C, the circumference of the circle, but in fact there is no polygon with altitude R and circumference C.

This picture is the basis, no doubt, for Dante's statement in *The Banquet* that a circle cannot be measured exactly "because of its arc." The polygon has straight sides, by definition, and no matter how many refinements one makes, the sides will still be straight, although there

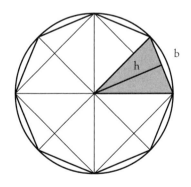

Area of polygon = $(1/2)\, h \times$ circumference

The area of $1/N$th of the N-gon is $(1/2)hb$, where h is the height of one wedge of the N-gon, and b is its base. The area of the N-gon is therefore $(1/2)h \times Nb$.

will be more of them. They will never become curved like the circle. And the area of the polygon will never actually be the area of the circle, no matter how many sides it has.

Now comes the peculiar Archimedean genius that Plutarch describes so well, an idea so simple that afterward we imagine that we might have thought of it ourselves. Archimedes uses this picture to prove that a circle with radius R and circumference C has area $(1/2)$ RC, exactly the area the polygon *would* have if it could turn into a circle, with h, the altitude of the polygon, becoming R, the radius of the circle, and Nh, the circumference of the polygon, becoming C, the circumference of the circle. (Note that this is just the usual formula for the area of a circle, because the circumference of a circle is $C = 2\pi R$.) Archimedes represents this area $(1/2)RC$ as the area of a right triangle with height R and base C, as shown below.

Archimedes proves that the $(1/2)RC$ triangle *cannot be less* than the circle, and also that it *cannot be greater* than the circle. Therefore it must *be* the circle, and the formula is true, as if the polygon had somehow *become* the circle.

We will show only that the triangle cannot be less than the circle. The proof that it cannot be greater is similar. Therefore, assume that the triangle is less than the circle, and show that this assumption leads

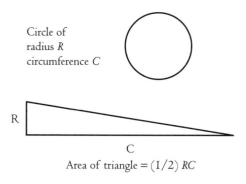

Circle of
radius R
circumference C

R

C
Area of triangle $= (1/2)\ RC$

The triangle has the same
area as the circle.

to an impossibility. Now the polygon, at each doubling of the sides, incorporates more than half the remaining area of the circle that wasn't already contained in it. Since, by our assumption, the triangle is less than the circle, the area $(1/2)h \times Nb$ of some polygon in the sequence will eventually exceed the area $(1/2)RC$ of the triangle, as indicated below.

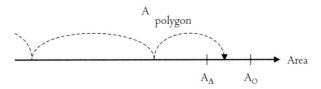

If $A_\Delta < A_O$, i.e., if the area of the triangle were less than the area of the circle, then in the sequence of polygons there would be one that has greater area than the triangle, an impossibility.

But this is impossible, because the altitude h of each triangular "wedge" of the polygon is less than the radius R, and the sum of the bases of these triangular wedges Nb is less than the circumference C, and therefore $(1/2)h \times Nb$ is *less* than $(1/2)RC$, not greater. Thus our assumption that the triangle is less than the circle was wrong, and the triangle is not less than the circle. In the same way, using polygons on the outside of the circle, Archimedes shows that the triangle also cannot be greater than the circle. Thus it must be equal, and the measure of the circle has been found: it is $(1/2)RC$.

The Archimedes Proof as Metaphor

If we look back at the last lines of *Paradiso* with the Archimedes proof in mind, we see a startling correspondence. The final lines read almost like a paraphrase of the proof. Let us look at this correspondence step by step.

The first step of the proof is to draw two perpendicular diameters in the circle, the sign of a cross, as in the first figure. This appearance of the cross must have struck Dante forcibly when he studied the proof. Surely this step ought to be visible in *The Divine Comedy*: and it is. In the circle representing Christ, "I saw painted our own human form." That is what the cross is: the image of a human being (crucified), a man outstretched, and the beginning of a transformation that takes the straight lines, representing the human, or the measurable, to the curved, the divine, or unmeasurable, the mystery of Christ's death and resurrection. The odd word *painted* suggests the geometer's drawing tool.

Now Dante compares himself to the geometer who wants "to see how the human image conforms itself to the circle and finds its place there." In the proof this is just what happens. The cross gives rise to the square, and then to the sequence of polygons that conform themselves, with increasing accuracy, to the circle. The language of the human "conforming itself to the circle" has a very literal, geometrical meaning in the proof.

The geometer, however, sees that the polygons never actually become the circle. No matter how far one goes in the sequence, the polygon always has straight sides, and the circle has a curved arc, as Dante said in *The Banquet* when he emphasized that any round thing is unmeasurable because of its arc. At this point the geometer is trying to understand how the polygon, in an infinite process, somehow becomes the circle.

With a flash of insight, one that has no simple picture to go with it but is clear to the intellect, the geometer proves that the polygon *does* become the circle, the "wheel that is turned evenly" (that is to say, unlike a polygon, which would be an uneven wheel). This is the end of the proof. The flash of lightning is a metaphor for the genius of Archimedes, which in turn is a metaphor for divine understanding.

Thus the Archimedes proof functions as an extended metaphor for the union of the finite with the infinite, the human with the divine. In doing this, Dante once again shows a mathematical sophistication that is completely unexpected. This is exactly how modern mathematics deals with the infinite. As Archimedes first showed, our minds can comprehend and even use the infinite, by methods that involve only finite things. In the proof of the area of the circle, although we imagined an infinite sequence of polygons, we only had to use a finite number. We only had to go far enough in the sequence to find one that had a large enough area, and this happens at some finite step, as in the last figure of the Archimedes proof. And yet, since the area of the polygons becomes the area of the circle, the polygons must *become* the circle. In particular, we see that the limit of an infinite sequence may have new properties that no individual member of the sequence possessed (here, a curved side, when all the members of the sequence had straight sides). Dante must have understood this perfectly. Otherwise the metaphor could not have occurred to him.

When did Dante conceive of these mathematical images? It is known that Dante broke off and never returned to writing *The Banquet*, his book of classical learning, in order to begin writing *The Divine Comedy*. It is not impossible that it was the discovery of this mathematics that gave Dante the compelling structure for the end of the poem and made it impossible not to embark on his epic journey.[3] That is, we may owe the very existence of *The Divine Comedy* to geometry, because he really did know where he was going.

The unexpected appearance of Hellenistic mathematics at the climax of *The Divine Comedy* is a crucial datum for understanding mathematics in the Renaissance. In the first place, it lets us know that the full sophistication of classical mathematics was available as early as Dante's day, 300 years before Galileo. Dante's understanding of Archimedes was not superficial. The concept of limit, the concept that Dante was using, was introduced as the new concept in calculus, invented by

Newton and Leibniz some fifty years after Galileo. Dante, with a remarkable sense of what is important, seized on this idea as the single most interesting concept in classical mathematics. We are left with the impression that if thirteenth-century mathematical culture had been different, Dante might be known to us as the great mathematician of his age.

In the second place, we see how precariously confined mathematics was to an obscure and recondite corner of philosophy. Ptolemy had sandwiched mathematics between physics and theology as the science of the Heavens. We know that eventually, in the work of Galileo, mathematics broke out of this confinement into physics, becoming also the science of the Earth. But in Dante it broke out in the other direction, indicating that the meaning of mathematics was very much up for grabs.

Commentaries

We can look back, to some extent, to see how these lines were understood in Dante's own time. The tradition of annotating the *Divine Comedy* began very early, and from 1375, some fifty years after Dante's death, we have the commentary of Benvenuto da Imola. On the final lines, Benvenuto writes, "He explains his highest effort with a most elegant simile of a geometer, who, wanting to measure the circle, devotes himself to it completely; and however much the author is seen to speak generally about geometry, this instance, which he places most highly, is proved by the philosopher Archimedes; about whom it is well known, as Livy and others write . . ."[4] Thus Benvenuto seems to know that this passage is not just another general allusion to geometry, but that it refers to a specific proof by Archimedes. He knows this, apparently, only through some sort of tradition, and doesn't understand it himself, because the reference he gives is to the histories, which tell the story of the death of Archimedes at the hands of a Roman soldier, but say nothing about any proof. The reference should have been to Archimedes's treatise. Apparently a tradition that correctly understood the end of the poem was beginning to be lost.

Perhaps misled by Benvenuto, subsequent commentators seem to have thought that the poem is referring here to the death of Archimedes, and one (Chiose Vernon, 1390) injects the whole story of the siege of Syracuse at this inopportune moment. The interpretation now seems to be that just as Archimedes was staring fixedly at the circles at the moment of his death, and was oblivious to everything else, with just such concentration was Dante staring at the vision of the trinity.

This bizarre interpretation of the climax of *The Divine Comedy*, in which we must imagine a geometer about to be killed, could not be maintained. It vanishes after the 1400s, but with it goes all mention of Archimedes, a case, perhaps, of baby and bathwater, since Archimedes was part of the metaphor, just not in this way. After the 1400s no commentator on this passage ever mentions Archimedes by name again. Thus part of the intended meaning in the climax of *The Divine Comedy* has actually been lost. The end of *The Divine Comedy*, instead of being read as a triumph of the human intellect, has come to be read in just the opposite way, as submission to a mystery that is beyond us. Knowing the mathematics actually changes the meaning of the poem.

Painting

The artist Albrecht Dürer tells us how he first learned of the geometrical theory of painting, a problem that occupied him for the rest of his life. It happened in his own city of Nürnberg in 1500 in the company of an Italian visitor "named Jacopo, born in Venice, a good artist, who showed me a man and a woman that he had made in true proportion, so that I had rather have seen his method than a new kingdom, and if I knew it, I would have it printed to honor him, for the general benefit of all. I was still young then, and had never heard of such things. I was in love with art, and I set myself to find out how he did it. But this Jacopo didn't want to show me, I saw that very well."[1]

These mathematical methods, for a while at least, were secret, although they would be published eventually. Dürer published his own self-taught version in 1525. But six years after the meeting described above, being now in Venice, he still had to go to considerable trouble to find out anything. In letters to his friend and patron Willibald Pirckheimer he frets over the jealousy of the younger Venetian artists, from whom he expects nothing, but he also describes his good relationship with Giovanni Bellini, who commissioned a painting from him, and of whom he writes, "He is very old, and still he is the best painter of them all." Bellini could have initiated Dürer into the methods of perspective painting, or found someone else to do it, but it was apparently not so easy. At the end of his stay Dürer writes, "I shall have finished here in ten days; after that I should like to ride to Bologna to learn the secrets of the art of perspective, which a man is willing to teach me."[2] So after spending most of a year in one of the great centers of Italian painting, he was now planning to try someplace else.

Dürer and his contemporaries believed that the Greeks had known how to paint with perfect realism, but that this secret had been lost. The rivalry among artists in the Renaissance to rediscover the ancient secret seemed to echo the very stories that convinced them that there was such a secret in the first place, stories like one told by Luca Pacioli in his 1509 book *De Divina Proportione,*

> As one reads in Pliny's *De Picturis,* there was a painting competition, in which Zeuxis and Parosasius challenged each other with their brushes, and Zeuxis painted a basket of grapes, surrounded by grape leaves. He placed his painting out for all to see, and the birds came to it as if they had been real grapes, to eat them. And Parosasius painted a veil. At this, when Parosasius had set up his picture, Zeuxis said, thinking it was a real veil covering the picture that he had made for the competition, "Take the veil off, and let everyone look at your picture, as I have done with mine." And this was how he lost! Because even if he had fooled the birds, which after all lack the gift of reason, his opponent had fooled not just a man, but a master of the art.

Unfortunately there is nothing in the story to say how the effect was achieved. Perhaps the closest thing to a technical description of a method comes from a very odd place, the preamble to the seventh book on architecture of Vitruvius, describing scenery for a play:

Agatharcus, in Athens, when Aeschylus was bringing out a trag-
edy, painted a scene, and left a commentary about it. This led
Democritus and Anaxagoras to write on the same subject, showing how, given a center in a definite place, the lines should naturally correspond with due regard to the point of sight and the
divergence of the visual rays, so that by this deception a faithful
representation of the appearance of buildings might be given in
painted scenery, and so that, though all is drawn on a vertical
flat facade, some parts may seem to be withdrawing into the
background, and others to be standing out in front.[3]

Vitruvius certainly does not understand the perspective method, but
his description of it, especially given that he is writing some 400 years
after the event, is amazingly accurate. Also surprising is the early date
that Vitruvius gives for this feat—well before Plato, and only a few
generations after Pythagoras. There is no corroboration for what Vitruvius says. The commentaries of Agatharcus, Democritus, and Anaxagoras on perspective have not come down to us, nor do any ancient paintings give unambiguous evidence for the perspective construction.
Renaissance commentators on this passage had nothing more to say.

The effort to recover the methods of the Greeks, whatever they
may have been, had a much longer history in Italy than in Germany, as
Dürer was ruefully aware. Even in the 1300s, in the paintings of
Giotto, one finds some of the features that would later be used systematically to achieve a realistic illusion of depth and three-dimensionality.
As early as 1425, the Trinity fresco of Masaccio was so accurately
worked out by geometry that although it was a painting on a wall,
people said that it was as if the wall had been pierced. By the late
1400s the geometrical method behind such paintings was being described in handwritten books (discretely circulated, we must assume).

The theory of perspective painting is the application of Euclidean geometry to the space around us. The success of this model of space in the art of painting, and in representing spatial relationships more generally, is perhaps the reason why Euclidean geometry seems like common sense to us, even while it didn't seem like common sense to Dante, who lived only a few generations before the perspective revolution in painting and who knew Euclid very well.

5   The Power of the Lines

The few surviving frescoes of the ancient world do not tell a clear story about whether the Greek and Roman painters were trying to create realistic illusions or even whether they would have known how. Realism is just one possible aim for a painting. After a period of infatuation with geometrical methods in the Renaissance, European painters became tired of this idea. Islamic painters, who had access to all the same geometry as Europeans, never became interested in it at all.

Still, it is common for paintings in many traditions and cultures to show more distant objects smaller, thus giving a rough indication of depth, the third dimension. The perspective construction in painting just makes this idea precise.

The Visual Cone

When Italian painters in the 1400s made it an explicit goal to construct paintings realistically and accurately, it did not suffice, at least not for long, to find rules of thumb that seemed to work. Many of these rules turned out to be wrong. The humanist and architect Leon

Battista Alberti, who was the first to write about the perspective technique in *De Pictura,* 1435, describes how one could tell that a rule commonly used in his day was incorrect.

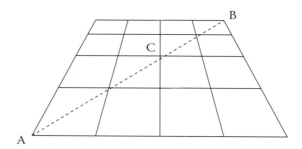

In this *pavimento* the straight line from corner *A* to corner *B* should go through corner *C,* but it misses. Thus the rule for constructing it must have been incorrect.

The figure shows a *pavimento,* that is, a horizontal floor paved with rectangular tiles. The impression that the tiles are receding away from us is indicated by their decreasing size. The rule used to construct it is this: as the tiles recede, they are reduced in height at each step by the factor $2/3$. The resulting floor looks like a possible rendering of what we might see, but a simple geometrical check shows that there is a mistake. Alberti points out that the diagonal through the floor should hit the corners of the tiles, but it doesn't.

It is not enough that a painting should just look right, as this one does. It should be literally and perfectly correct. This is the standard that Alberti desires. It is not completely clear that Alberti himself knew a correct construction when he wrote this, although he had a sharp eye for incorrect ones. What he says stops short of a fully explicit method, but he does state the problem.

According to Alberti, the basis of the construction is the *visual cone* (or *visual pyramid*), the cone of rays emanating from the eye to the objects that the eye sees. When we look at an object, the object itself is the base of the cone and the eye is the vertex, as shown below.

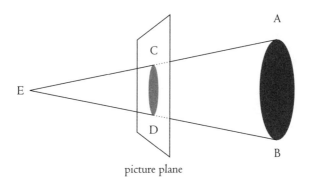

picture plane

If the eye is at *E*, the object *AB* is represented in the picture by the section *CD*. A painting of *AB* is thus a cross-section of the visual cone with vertex *E* and base *AB*.

A painting is a cross-section of this cone. It can also be thought of as a "window" through which one looks out at the object. It is apparent in the figure that the construction of the painting depends crucially on the assumed placement of the eye, point *E*. It is only possible to say what the picture should be after the point *E* is chosen. Also, the finished painting will have to be looked at from just this point to be seen in the way that is intended. Otherwise we will see things out of place. This requirement on the placement of the eye seems so strict that it is not clear that the construction will succeed in creating the illusion of realism, even if it is carried out perfectly. Another potential difficulty is that we have two eyes, not just one, and we get some of our sense of depth and three-dimensionality from this binocular vision, not from just the information available to a single eye. The theory completely ignores this effect.

Remarkably, the objections of the last paragraph, while worrisome in principle, are not so serious in practice. Something about the psychology of vision makes us quite tolerant about seeing a perspective painting from the wrong place. We still get the effect. Binocular vision also does not ruin it. In spite of weaknesses in the concept of the perspective construction, it basically works.

The visual cone is the perspective theory in essence, but it shows us almost nothing that would be helpful in actually carrying out the idea. Suppose the object were a paved floor, for example. On the basis of this theory, what should the painter put onto the canvas? The relationship between the object and the painting is geometrically complicated, even if we understand it in principle. The theory doesn't yet tell us what to do.

The problem of turning the visual cone into a useful procedure for artists was probably solved more than once. Masaccio's frescoes seem to be correctly constructed.[1] Filippo Brunelleschi, the architect of the dome of Santa Maria del Fiore, Florence's great cathedral, famously demonstrated the correctness of his painting of the adjoining Baptistery with a mirror device that allowed the viewer to compare the painting with the real thing, and to verify that they were geometrically identical. Neither man left a written account of how he did it, though, and so we cannot be sure that either had a real solution. The first complete description of the perspective construction in writing, by Piero della Francesca, makes essential use of ideas from Euclid's *Optics* and Euclid's *Elements*. Nothing short of this level of sophistication, no plausible procedure leading to apparently correct paintings, would have been sufficient for a real proof of correctness. One gets the feeling that at least some of the principals in this endeavor wanted to be absolutely sure that it was right. That proof required mathematics at the level of the Hellenistic Age.

Euclid's Optics

The idea of the visual cone comes directly from Euclid's *Optics.* Euclid talks about the rays of the cone as if they were actual emanations from the eye, a kind of sensory apparatus that reaches out from the eye and measures things, like a geometer's compass. The measure of something, that is, its apparent size, is just the angle it makes at *E,* the eye. This is Euclid's key insight. The apparent size of something is really its *angular* size. Things that are far away look smaller because their visual cone makes a smaller angle at the eye. A painting is correct if the object (*AB* in the preceding figure of the visual cone) and its representation *CD* make the same angle at the eye.

It is worth noticing how stark and simple this mathematical model is, and in particular how much it leaves out. Later treatises on optics, under the influence of Aristotle, couldn't resist putting in much more about real vision. The result is that they became essentially useless. The earlier, simpler *Optics* provided the basis for perspective. Ptolemy also wrote an *Optics* in five books of which Book I and much of Book 5 are lost. We do not have the philosophical introduction that was undoubtedly much of Book I, but Book 2, where we might expect to get down to business, is verbose, classifying all the things you might want to look at in terms of density, rarefaction, color, and so on; things that go unmentioned in Euclid and that seem irrelevant. When Ptolemy finally does get to geometry, he first takes up binocular vision, and in particular the way nearby images seem to jump if you alternately cover your right eye, then your left. Only after all of this does he take up Euclid's simple starting point.

The Arabic *Kitāb al-Manāzir,* a large and impressive work on optics of Ibn al-Haytham of around 1030, was translated into Latin around 1220 as *De Aspectibus.* It is competent and original, arguing many points with Euclid and Ptolemy, but it ultimately takes its general approach from Ptolemy, to its detriment. Ibn al-Haytham's work, even more than Ptolemy's, demonstrates how Aristotelian philosophy, in its eagerness to include everything and classify everything, obscures the clean insight that geometry could bring to the problem of vision. Even Euclid's basic insight that visual size is *angular* size is obscured in al-Haytham. Ptolemy has already pointed out, in a tentative way, that our judgment of the size of things makes use of more than just their angular size, but he only says this as an aside. In al-Haytham the point is argued at length. When an object moves to a greater distance, we do not perceive it as getting smaller, he says, even though its form occupies an apparently smaller space. That is, we mentally make corrections for distance, and we use what we know about the permanence of things to interpret visual information. This is true, of course. But the effect is to entangle the simple geometrical theory, partial though it is, with a more complete theory that cannot succeed because of its complexity, because it tries to explain everything, even perceptual psychology.

Euclid simply sidesteps all the complexity. Here is how the *Optics* begins: First, let it be assumed that lines drawn directly from the eye pass through a space of great extent; second, that the form of the space included within our vision is a cone, with its apex in the eye and its base at the limits of our vision; third, that those things upon which the vision falls are seen, and that those things upon which the vision does not fall are not seen; and fourth, that those things seen with a larger angle appear larger, and those seen within a smaller angle appear smaller, and those seen within equal angles appear to be of the same size.[2]

One of the first propositions Euclid proves is that parallel lines receding from us do not *look* parallel. This is a familiar fact, perhaps best seen when looking down a long straight section of railroad track at the parallel rails. They seem to come together as they recede, and would even seem to meet at the horizon if one could see that far. Euclid's proof uses the following figure, which represents the viewer's eye at point *E.* The receding parallel lines are *AB* and *GD.* The distances *TK,* *ZL,* and *BD* are all equal because the lines are parallel, but they don't *look* equal, because the angles they make at *E* are smaller the farther away they are. Thus the farther ones look smaller, and the parallel lines seem to come together.

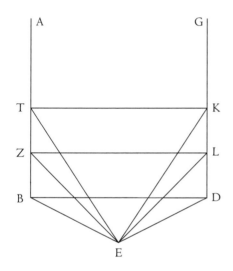

The parallel lines *AB* and *GD* appear to approach each other because the equal distances between them, *BD, ZL,* and *TK,* appear successively smaller, subtending smaller angles at *E,* the eye.

This is how the *Optics* proceeds, diagramming situations and pointing out the angles at *E,* the eye. It describes situations, like the one above, where things that are the same look different, and other situations where things that are different look the same.

The last proposition of Euclid's *Optics* seems to be on the verge of something important, but inexplicably it stops. It says, "If from the meeting point of the diameters of a square a perpendicular is drawn, and the eye rests upon this, the sides of the square will appear equal and the diameters also will appear equal." That is, if you look down on a horizontal square from a point above its center, then it will look square. This is obvious by the symmetry of the situation, and it is easy to prove, as Euclid does. But what about the case where you look from some other place, not from above the center, but from off to the side? How would it look then? Not square. This is exactly the problem of representing a square floor, or a *pavimento,* seen from a general position. It is the basic problem of the perspective construction to find how the square looks in this case. Euclid leads up to it and stops.

Piero della Francesca

The quattrocento painter Piero della Francesca (c. 1410–1492) is a magisterial personality. His artistic reputation, high enough in his own day, waned over the centuries until his rediscovery in the early twentieth century, but in our time he has become almost a cult figure, admired above all his contemporaries for the uncanny "stillness" of his figures and the enigmatic quality of his paintings. Among his best-known works are two in the National Gallery in London, the *Baptism of Christ* and the *Nativity of Christ,* the fresco cycle *Legend of the True Cross* in Arezzo, and the exquisite *Flagellation of Christ* in Urbino. The 500th anniversary of his death, in 1992, was the occasion for symposia and for projects to bring out new editions of his books, for he was an author as well as a painter.

And he was a mathematician. His formal schooling in mathematics could not have gone beyond the abacus school in his birthplace of Borgo San Sepolcro (now Sansepolcro), and there is some doubt about that because no evidence exists that San Sepolcro even had an abacus school.[3] Perhaps he studied a textbook on his own. As the son of a leather tanner and shoemaker he would have been expected to learn commercial arithmetic. However it happened, Piero's mathematical ability showed itself very early, according to the biographer Giorgio

Vasari in his *Lives of the Painters*. A habit of teaching himself is perhaps the only explanation for how he was able to learn not just the practical arithmetic of the abacus school but also Euclidean geometry and even Archimedes.

How would a young man of the artisan class get access to the texts of Hellenistic mathematics? On this point Piero had very good luck. He was employed for a time by Duke Federigo da Montefeltro of Urbino, for whom he made many pictures, as he himself says later on. Federigo was a consummate Renaissance nobleman, skilled in arms and diplomacy, and a great patron of learning. Although his tiny dukedom never attained anything like the wealth or power of better-known states like Florence, its palace and its court were universally admired. Above all, its great library seems to have been, at least for awhile, the most complete in Europe. Piero had access to this library.[4]

A contemporary, Vespasiano Bisticci, describes it:

> He alone [Federigo] had a mind to do what no one had done for a thousand years or more; that is, to create the finest library since ancient times. He spared neither cost nor labor, and when he knew of a fine book, whether in Italy or not, he would send for it . . . beginning with the Latin poets, with any comments on the same . . . next the orators . . . and all Latin writers and grammarians of merit; so that not one of the leading writers in this faculty should be wanted . . . As to the sacred Doctors in Latin, he had the works of all four, and what a noble set of letters we have here; bought without regard of cost . . . Likewise all the writers on astrology, geometry, arithmetic, architecture, and *De re militari* . . . The Duke, having completed this noble work at the great cost of thirty thousand ducats besides the many other excellent provisions that he made, determined to give every writer a worthy finish by binding his work in scarlet and silver.[5]

This loving description of the books themselves, together with their cost, is partially explained by the circumstance that Bisticci was a Florentine bookseller and Federigo's supplier. Needless to say, the books were manuscript books. Printed books hardly existed at this time, and they would have been considered cheap and second-rate.

Piero did not merely use the library at Urbino, he contributed to it. His pathbreaking manuscript book *De Prospectiva Pingendi* (On Perspective Painting) found a home there, although it did not get the scarlet-and-silver treatment. This volume now resides in the Vatican Library along with the rest of Federigo's collection. It is an unreadable masterpiece.

De Prospectiva Pingendi is not meant to be read, but to be worked through. Opening it at random, we are almost certain to find something like "First draw .1. intersecting .BD. in point .31., and draw .2. intersecting .BD. in point .32, and draw .3. intersecting .BD. in point .33., and draw .4. intersecting .BD. in point .34., and draw . . ." It reads like a computer program, written in a language that doesn't have a command to say "repeat until done." It consists of precise instructions specifying explicitly every point and line in a series of perspective drawings. Simply to read it makes no sense.

What a surprise, then, to follow these instructions, using the book as it was intended to be used, and to find that it is actually fun. The painstaking repetition, describing every line and point, is not tedious now, but reassuring. You feel confident as you follow these systematic directions that your work is entirely correct. The drawings are organized like theorems in Euclid, one after another, building upon each other, and gradually increasing in sophistication. Each one teaches you something or establishes something that will be used again.

Over the next two centuries, and especially with the advent of printing, other perspective books were published, and if they referred to Piero's book at all, it was often in a disparaging tone. They all followed its pedagogical plan very closely, though.[6] And none of them could diminish the greatest achievement of *De Prospectiva Pingendi:* the proof by Euclidean geometry of the correctness of the perspective construction, making perspective painting a new science.

The Correctness Proof

It is certain that Piero worked out the first section of *De Prospectiva Pingendi* in the library at Urbino, because he frequently cites specific propositions from Euclid's *Elements* and Euclid's *Optics.* He must have had these books in front of him. The correctness proof is more than a mathematical proof, it is also a drawing, the basic starting point for all the drawings that follow. Only when this foundation has been laid do the drawing exercises commence.

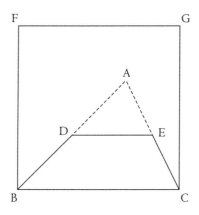

The vertical square *BCGF*, if it falls backward to lie flat, appears as the foreshortened figure *BCED.* The point *A* is the vanishing point for horizontal lines that recede perpendicularly into the page.

The first figure shows the basic ingredients of the perspective construction. Imagine that we are standing and looking perpendicularly at a square wall *BCGF* in front of us. The wall is then allowed to fall slowly backward until it is lying flat, as a square floor. This floor *BCED* is constructed by locating the "vanishing point" *A*, at the level of the eye, and drawing the two lines *AB* and *AC.* The two edges of the square floor that recede from us would appear to meet at the vanishing point *A* if they were prolonged, so they must lie along these two lines. The construction is finished by locating the back edge of the square floor *DE*. This basic idea was common practice in the 1400s. Indeed, the whole construction may have been in use long before it was actually proved to be correct.

Piero's proof connects the figure of the visual cone to the figure of the square floor.[7] The picture plane is taken to be the original standing square *BCGF*, so that we may reinterpret this figure as a view through a square window *BCGF* of a square floor *BCED*.

A second figure shows the situation from the side. An observing eye at *I* looks through the picture plane at the square floor, represented by the length *BC* in the figure. The visual cone is therefore *BIC*, and its section in the picture plane is *BJ*, the part of the painting that will represent the floor.

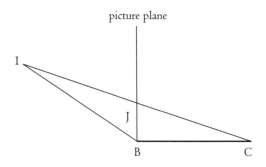

An observing eye *I* looks at the square floor *BC*, in a view from the side, through the picture plane, where *BC* seems to occupy the region *BJ*. The visual cone is *BIC*, with vertex at *I*.

A third figure shows the situation from the top. The square floor, now labeled *BCGF*, is seen from above, and the eye at *I* sees *CG*, the back edge of the floor. By sectioning the visual cone *CIG* with the picture plane *BF*, we see that *CG*, the back edge of the square, should be represented by the distance *JK* in the painting. We get the same distance *JK* no matter how we position *I*, as long as we keep the same distance from the picture plane, so we might as well take *I* in the same place relative to *BC* as it was in the side view.

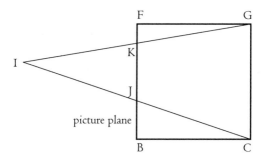

An observing eye *I* views the square floor *BCGF* in a top view. The back edge of the floor, *CG*, is seen by the visual cone *CIG*. Taking *BF* to be the picture plane, we see that the section of the visual cone is *JK*. This length *JK* represents *CG* in the painting.

The distances *BJ* and *JK* are the data needed to represent the square floor. We notice that they are both available in the top view, even though the argument to justify *BJ* used the side view, not the top view.

The observation that the top view contains, in a sense, *both* the side view and the top view suggests Piero's final diagram, which contains side, top, and front views of the floor and the painting all at once.

The only good way to understand it is to realize that it contains the three previous diagrams. Looked at in this way it is possible to interpret it piece by piece, and Piero now proves that the pieces fit together consistently. In particular the back of the floor, represented by *DE* in the painting, is just where it should be according to the side view, namely at the level of point *J*. And, quite amazingly, *DE* has the length

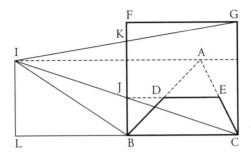

This diagram incorporates the three preceding figures. Piero's proof that the method for constructing *BCED* is correct is to note that the back of the square floor, represented by *DE*, is at the right level, namely the level of point *J*, and has the right length, namely the length *JK*. That last assertion, *DE = JK*, is not so obvious.

required by the top view, namely *DE = JK*. That is the only tricky part of the proof. The length *DE* comes out right *only* if the vanishing point *A* is taken at the level of the eye *I*, as shown in the last figure here. Thus Piero does not assume the existence of a vanishing point (and he never calls it that). Rather he proves its existence. Piero proves that *DE = JK* by a clever use of similar triangles.[8]

The Drawing Exercises

The square floor *BCED* is still only a blank region in the painting, but it defines the visual space. Everything else will be located with reference to it. To appreciate Piero's emotional commitment to this system, it is worthwhile seeing a few more of its details and noting how beautifully and simply they work to solve the problem of representing real things. Piero shows in a series of drawing exercises how to construct the appearances of figures in the square floor.

Piero's word for the perspective appearance of things is "degraded." Things have their "proper form" and they have their "degraded form," the form they take in the painting. The degraded form of a set of horizontal lines receding orthogonally away from the picture plane is a set of lines intersecting at the vanishing point *A*, for example. This is the lesson of the first drawing exercise, below. Subsequent exercises add sophistication, a little at a time. One gets the impression that Piero is a very thoughtful teacher.

The first drawing exercise is to divide the square floor *BCED* into equal "strips." The method is to divide the front edge *BC* into equal intervals by adding points *F, G, H, I* and to connect these points with the vanishing point *A*. This is correct, because by similar triangles the back edge *DE* is now also divided into equal parts, as Piero has already proved in an introductory section. This amounts to a proof that *A* really does have the property of a vanishing point, a point where all parallel lines seem to converge in the painting.

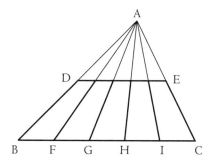

Dividing *BC* into equal pieces and connecting to *A* divides the "degraded" square *BCED* into equal strips.

Alberti had suggested using the diagonal of a square to check the correctness of a *pavimento*. For the next exercise, Piero turns this idea around and makes it the basis of his method for placing the transversals of the *pavimento*. They satisfy Alberti's criterion by construction. The transversals, parallel to *BC*, must be located where the diagonal *BE* intersects the orthogonals *FA, GA, HA*, and *IA* constructed in the previous exercise, as shown next.

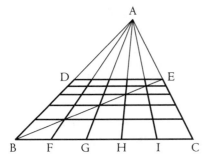

The transversals of a square *pavimento* are determined by the diagonal.

Thus far we have constructed a degraded square and a degraded version of the square tiled by smaller squares. Piero's general method begins to be clear in the following exercise, shown in the figure below, where he degrades an octagon .fghijklm. He begins with a proper square and a proper octagon. The process of transforming involves

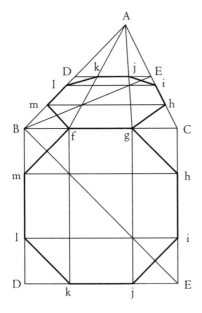

An octagon .fghijklm. is shown in its proper form in the square below, and in its degraded form in the degraded square above. Note how the vertices of the octagon are found as the intersections of constructed lines. This construction is essentially the same as a *pavimento* with tiles that are not all square.

drawing vertical lines from the vertices of the proper octagon up to
BC, then continuing them to *A*. Intersections with the diagonal have
the same significance as in the previous exercise, since they are essentially locating the transversals, parallel to *BC*, of an irregular tiling of
the square.

The general method for degrading any figure in the plane is illustrated in the next exercise. The vertices of a triangle .fgh. in proper
form are transformed into the vertices of the degraded triangle by
drawing two paths for each vertex. The first goes vertically to *BC* and
continues on to *A*. The second goes horizontally to the diagonal, vertically to *BC*, on toward *A* as far as the degraded diagonal, and then
horizontally to intersect the first path. This intersection is the location
of the degraded vertex. In the figure the vertex *h* is a particularly clear
example.

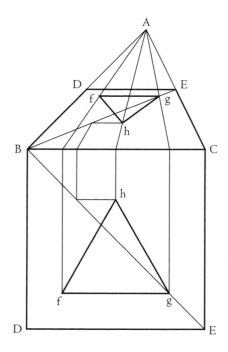

A triangle .fgh. is shown in
proper form in the lower,
proper square, and the
degraded form is constructed
in the degraded square.

We notice that the transformation inverts the triangle. Thus objects in their true form must be represented upside down if the degraded form is to come out right side up.

Adding the Third Dimension

Book I of *De Prospectiva Pingendi* is all about two-dimensional figures in the square floor and how to degrade them. Book 2 introduces the third dimension and solid objects rising vertically over the ground plans in Book I, like houses and wells. The constructions get correspondingly more complex. Now they produce degraded versions of three-dimensional objects.

In Book 3 Piero considers truly complicated objects, like capitals on columns and human heads. Interestingly, he even geometrizes these, suggesting that one could mark many points on a head, enough to determine it accurately, and degrade these selected points by geometry. It amounts to thinking of a head as a complicated polyhedron. The data that would go into this process would be frontal, side, top, and bottom views of the head, with each marked point appearing in more than one view. Amazingly, Piero actually seems to have made such a visual database of views of heads. In *De Prospectiva Pingendi* he describes how to transform such data to create angled views of the heads. The process, once again point by point and line by line, goes on for pages and pages. Anyone who thinks Renaissance artists could not have had the patience to do such things should look at *De Prospectiva Pingendi*. Whatever patience it took to carry out this process, it took at least as much patience to write it all down in such detail.

The method for transforming things from their proper form to their degraded form in Book 3 is slightly different from the earlier books.[9] Piero says that when you have so many points to transform, there are just too many construction lines if you use the method of

Books 1 and 2. Rather he suggests using a horsehair or silk thread, stretched from point to point, to replace the earlier construction lines drawn on the surface. To record intersections of lines he describes a system of little rulers, made of thin, straight strips of paper and wood, on which you mark positions. With this method the surface remains uncluttered because the construction lines largely disappear when the horsehair and the rulers are removed. The geometrical method is still the same, though, and it still begins with views in "proper form." The database of heads would contain views in profile ("one eye"), face-on ("two eyes"), and from directly above. These data would then be transformed by geometry. Perhaps, one can't help thinking, this is why Piero's human figures look so "still." They were never real to begin with. They are pictures of pictures.

The Consequences of Correctness

Piero, because of his proof, knew that the perspective construction was not just a plausible idea that seemed to work. It was, for what it aimed to do, absolutely correct and perfect. As a consequence, Piero's commitment to geometry is complete. He seems to regard even the human form as a locus of points, ready for transformation.

Just imagine his impatience, then, with painters who had reservations about it. To be sure there were some like his contemporary Paolo Uccello, who, when called to bed by his wife, replied, "Ah, but it is so sweet, this perspective!"[10] But there were others who simply weren't buying it.

Book 3 of *De Prospectiva Pingendi* begins, "Many painters disparage perspective, because they don't understand the power of the lines and the angles, by which every outline and feature is shown in proper proportion . . . For a painting is nothing other than a representation on a surface of objects transformed . . . and I say that perspective is necessary in order to distinguish all quantities proportionally, like a true science, accomplishing the transformation of every quantity by the power of the lines." Piero seems to suggest that doubters simply don't understand the method.

But Book I ends with a more significant challenge. "I will remove the error of some who, not being very skilled in this science, say that many times, when they degrade the plane into equal parts, the foreshortened parts become bigger than the ones that are not foreshortened . . . They doubt that perspective is a true science, judging falsely because of their ignorance." This objection does not seem to come from people who fail to understand the method, but rather from people who *do* understand it, and even use it, but have found something wrong with it. Piero too, it turns out, is troubled by this objection. He cannot believe geometry would mislead him, but he is clearly uncomfortable about it.

The problem is indicated in the figure, illustrating Piero's Proposition 30, where we see that the critics were right. Sometimes the perspective construction, which we have come to understand through examples as a kind of foreshortening, actually produces a result that is

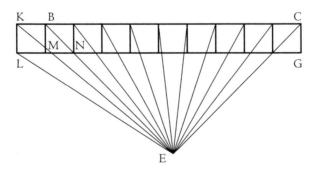

An eye at *E* sees *KB* with visual cone *KEB*. The section in the picture plane *LG* is *MN*, which is smaller than the object *KB*. But the object *KL*, with visual cone *KEL*, has section *LM* in the picture plane, larger than *KL* itself.

longer, not shorter, than the original object. The object *KL* shown here is represented by *LM* in the painting, which is longer than the object *KL*. This, they said, couldn't be right, and Piero reluctantly seems to agree.[11]

Piero sees that the problem occurs in the construction only for things placed rather oddly in the picture, way off to the side, at a viewing angle more than forty-five degrees away from the central position, so that the angle that the entire picture makes at the eye must be greater than ninety degrees. Piero somewhat lamely suggests that we don't see things so far off to the side. We should never have to stand so close to a picture that its edges are seen at such a large angle. His rule of thumb is that if the picture is seven units wide, the (centered) eye must be six units or more distant from the picture plane. Then, according to the approximate tradition of practical geometry, the angle at the eye would be only sixty degrees or less, and certainly less than ninety degrees.

This whole argument shows signs of having been thrown together hastily, and it is clearly unsatisfactory. In the first place, just by stretching our arms forward with an angle somewhat larger than ninety degrees between them, it is easy to see that we *do* see a wider angle in front of us than ninety degrees. Piero never addresses this observation. Instead he speculates on how the structure of the eye limits our field of view to ninety degrees, ascribing it somehow to the angle with which the optic nerve meets the crystalline humor in the eyeball. He worries that if we take our second eye into account the problem might resurface, and argues strenuously against this possibility. He even speculates on how animals see, since the problem shouldn't arise for them either.

The messiness of real phenomena, in this case the complexity of real vision, has intruded on the perfection of geometry. The problem is really internal to the theory. It is a problem of geometry, and not, as Piero would have it, just something that the unskilled do not understand. The only solution, in Piero's mind, was to go outside geometry, and here there was a wealth of possible remedies to be found in the elaborate Aristotelian theories of vision, from Ptolemy, to Ibn al-Haytham, and on to his Latin translators and imitators, but it was an awkward retreat from the clean geometrical theory that he had begun with. The argument in Proposition 30 even spoiled an interesting and original innovation in his Proposition 1. Where Euclid simply postulates that what we see as size is really *angular* size, Piero goes further, and justifies this idea by saying that the eye is, in effect, just a single point. Being structureless, the angle of the rays is all the eye has to work with. Now, in Proposition 30, Piero is forced to admit an elaborate structure for the eye after all. The whole theory has, implicitly at least, suddenly gotten a lot messier. Cavils of this kind are always in tension with mathematical theories, because mathematical theories are necessarily oversimplified. Piero's emotional reaction is clear, however: he believes in geometry, "the power of the lines."

The later history of this observation is amusing. Piero was right: one shouldn't use the perspective construction to degrade things seen at a grazing angle, with the eye close to the painting and the image off to the side. But the theory is also right: such constructions *will* be seen correctly if the eye is positioned in just the intended place. It is only that our tolerance for being in the wrong place is now rather small. The simplest case to think about is a picture of a sphere off to the side. According to the perspective construction, the section of the visual cone would be an ellipse, and it is true that if you paint an ellipse off to the side, it will appear foreshortened, and if you look from just the right place it will be foreshortened into a circle, the outline of a sphere—exactly as intended. But as a practical matter, it doesn't work to paint ellipses off to the side if what you are trying to represent is spheres. Seen from almost any place but the right place, they look like ellipses. You are better off painting circles, even though that is not what the perspective construction says, because the viewer will make the psychological adjustment and will understand what is intended.

The most extreme example of deliberately constructed perspective images—those intended to be seen from just the right place, looking at a grazing angle along the canvas—are called *anamorphic images,* and they became a kind of playful ingredient in a few paintings as the phenomenon came to be understood and exploited. The most famous example is probably the skull in Holbein's *The Ambassadors,* in the National Gallery in London. If you look in the normal way at this painting, you only gradually become aware of a mysterious blotch at the bottom of the painting in the center, with an orientation that goes from upper right to lower left. If you walk to the right, quite close to the painting, and look along the blotch down and to the left, a skull appears.

Piero himself played tricks with perspective. As a rule, when something can be painted accurately in perspective, he seems to feel it would be wrong not to do so. Even frescoes high up on a wall, not very visible, are constructed with as much care as everything else (but not to accommodate a viewer down below—that would be anamorphic).[12] Within the category of carefully constructed paintings, though, Piero sometimes makes interesting choices.

One striking example is *The Resurrection,* still to be seen in Piero's hometown, San Sepolcro. Four soldiers are sleeping, leaning up against the sarcophagus, or sprawling into the foreground. Their heads are tilted in ways that suggest the constructions of *De Prospectiva Pingendi,* Book 3. Towering over them is the risen Christ stepping out of the tomb. He stares straight out of the painting. There is something riveting and disorienting in the figure, going even beyond the drama of the moment. Attention to the perspective construction goes some way to explain the eerie effect. From the way we see under the soldiers' chins and do not see the top of the sarcophagus, it is clear, from the lower part of the painting at least, that we are looking from a very low point, definitely below the sarcophagus lid. The figure of Christ, however, is seen face on, even though He is looking horizontally at a level some four feet higher than the sarcophagus lid. Piero has put two inconsistent perspective constructions into one painting, clearly with intention.

This ambiguity in the perspective is what brings us face to face with Christ in such a shocking way. Piero has used the opportunity to play with naive realism and to suggest alternative realities.[13]

A second, much discussed, and very enigmatic example is *The Flagellation of Christ* in Urbino. This little panel is constructed with such detail and accuracy that everything in it can be located unambiguously in three-dimensional space. The vanishing point, though, is in a rather odd location. It is centered in the painting, but only about eighteen inches off the floor, and in a completely insignificant location, on a door jamb. The scene is presented as if from the perspective of someone who is kneeling on the ground. It seems unlikely that this is just a routine expression of humility, although great events are taking place before us. A conclave of three oddly dressed and seemingly unrelated men is taking place in the foreground, and in a background corner Christ is being flogged by a Roman soldier. One intriguing suggestion for the painting is that it was part of a diplomatic offensive from Cardinal Bessarion, who would then be one of the men in the foreground, to Federigo da Montefeltro, entreating him to join in a military expedition against the Turks.[14] In this case Piero could have encoded the message of supplication in the kneeling position of the constructing eye.

Perspective and Relativity

The perspective construction involves not just what is seen in the painting, but also something that is not seen, or at least not seen directly, namely the observer. A perspective painting, although it may seem to be "scientific" and objective, is necessarily also subjective, since it depends upon a particular point of view. As such it is more relativistic than absolute. The rules that determine the construction are strict, but they highlight the existence of multiple points of view, among which the artist must choose.

This awareness of multiple points of view, and the necessity of reconciling them, has become a principle of modern physics, most familiar through Einstein's relativity theory. It was already recognized and discussed by Galileo, though, in considering the possible motion of the Earth. The old argument for the stationary Earth, going back to Ptolemy, is essentially that if the Earth were moving, we would notice it. But Galileo suggests that the Earth could be moving after all, because if we are moving with it, then from our point of view the Earth is not moving, in the sense that it is not moving relative to us. It is a case of two points of view. In one of them the Earth seems to move, but in the other, that of someone moving along with the Earth, the Earth seems stable and at rest. Galileo never explicitly connects this idea of Galilean Relativity, as it is called, to the problem of multiple perspectives in painting, but just knowing the perspective theory, as he did, might have influenced his understanding of the somewhat subtle idea of multiple points of view.

Early perspective paintings do not use this freedom to express the presence of the observer. Instead a common device is to put the vanishing point in some place of significance *in the painting*, thus keeping everything objective, if artificial. The vanishing point in Masaccio's *The Tribute Money*, and also in Leonardo's *The Last Supper*, is in the head of Christ, as if to direct attention there, or to highlight Christ's importance. Piero is perhaps the first to use the vanishing point with subtler intention.

The New Geometry

In the poetry of Dante we notice the peculiar fact that, for Dante, geometry had a philosophical meaning that was not particularly spatial. Rather, geometry was a collection of true propositions. In the perspective treatises, beginning with *De Prospectiva Pingendi*, this rather "ungeometrical" interpretation of geometry cannot survive. The treatises are full of line drawings of real things, constructed by Euclidean geometry.

The practical proof of their correctness, requiring no effort at all to understand, is that they look right. The notion that Euclidean geometry consists of propositions that are true with certainty, while not contradicted, is eclipsed by the much more assertive idea that Euclidean geometry describes the world around us. If it doesn't, then how do you explain these pictures? Thus the perspective masters, and especially Piero with his unassailable proof of correctness, had pushed geometry back toward the meaning that it surely had originally, a model for spatial relationships.

The perspectivists also changed the capabilities of geometry by adding a powerful new tool, the painting as a transformation. Euclid's *Optics* did not have a very good way of talking about what we see, even though this was its ostensible subject. Just to say "the angle at the eye" is not a very complete description of what the eye sees. As a result, the *Optics* is not able to say things very precisely, and it suffers from the weakness of its tools. In the end, the *Optics* is not a very impressive book—it does not build from proposition to proposition as the *Elements* does. It seldom comes to any important conclusion, only that one thing appears bigger or smaller than another. Piero's treatise, by contrast, builds very systematically from beginning to end. What is more, the painting turns out to be a much more complete way to say what the eye sees. After all, when we see something, it has an angular size, to be sure, but it also has a shape and a location, geometrical characteristics that Euclid's *Optics* doesn't have a language to handle. Thus *De Prospectiva Pingendi* is more than an application of Euclidean geometry to a practical problem. It is the invention of a tool that extends geometry. Euclid's *Optics* ends just where the perspective construction should begin. In writing *De Prospectiva Pingendi*, Piero took up the job that Euclid's *Optics* had attempted and completed it. He may have aimed to equal the Greeks, but in the end he actually extended geometry in a new direction.

6 The Skin of the Lion

In 1545 the artist Giorgio Vasari published his *Lives of the Painters.* It is a collection of biographical sketches of the great artists of the Renaissance by one of their own, together with critical and appreciative comments on their works. It was highly popular in its day, and has remained an invaluable and entertaining resource ever since, in spite of its frequent unreliability. Vasari seems to be writing largely from memory or from hearsay. If he had known how future generations would pore over his every word, he might have felt obliged to check his facts more carefully, but it is a charming and useful book nonetheless.

One of the painters in Vasari's book is Piero della Francesca. Vasari is impressed by Piero's ability to foreshorten things accurately, and calls attention to numerous fine examples in Piero's paintings. Without mentioning *De Prospectiva Pingendi* by name he even says, "Piero gives us to know . . . how important it is to copy things as they are and always to take them from the true model; which he did so well that he enabled the moderns to attain, by following him, to that supreme perfection wherein art is seen in our own time."

Vasari's opening remarks, though, are not about Piero's art but about his mathematics, and they tell a sad story. Unhappy are those who labor for their own fame, only to have their work appropriated by another! This, says Vasari, is what has happened to Piero, and the plagiarist is Luca Pacioli, Piero's own countryman. "The very man who should have striven with all his might to increase the glory and fame of Piero, from whom he had learnt all that he knew, was impious and malignant enough to seek to blot out the name of his teacher, and to usurp for himself the honor that was due him, publishing under his own name all the labors of that good old man." In modern histories of mathematics Luca Pacioli is the most significant mathematician of the fifteenth century, but the best that Vasari can say of him is that he "sought to conceal his ass's hide under the glorious skin of a lion."

Luca Pacioli never wrote a perspective treatise, so Vasari cannot be suggesting that Pacioli plagiarized *De Prospectiva Pingendi*. Rather he refers to "many books" of mathematics written by Piero, and still preserved in Borgo San Sepolcro, the native town of both Pacioli and Piero. It is typical of Vasari that a few lines later on, without seeming to notice the contradiction, he says that these books are preserved in the ducal library in Urbino. Whatever the details, Vasari's sketch of Piero has much to say about otherwise unknown works of mathematics.

For centuries this account seemed to be an unfair libel of Luca Pacioli, who lectured and published energetically on mathematics until his death around 1514. He is sometimes called the "Father of Accounting" for advocating double-entry bookkeeping. His portrait, now to be seen in Naples, shows him in Franciscan habit, surrounded by tokens and symbols of his beloved mathematics. Why should Vasari attack him? Pacioli's brother Franciscans were inclined to defend him, and the citizens of Sansepolcro, despite possibly conflicting loyalties on the question, have put up a handsome statue of him. Vasari names the places where Piero's supposed mathematical works should be found, and this makes the accusation sound plausible, but as centuries passed and no one produced the books, it seemed to suggest the opposite, that they had never existed. Vasari's inconsistency in saying where to look was derided and his many other errors in matters of fact were called to witness how unreliable he was. What did an artist like Vasari know about mathematics anyway? Someone in Piero's family must have planted this outrageous story, which the credulous Vasari had accepted. Defenders of Pacioli pointed out that Pacioli, so far from being an enemy, lauds Piero highly, calling him *Monarca della Pittura*, Monarch of Painting, an epithet which has stuck.[1] (In retrospect Pacioli's praise looks carefully calculated to avoid mentioning Piero's mathematics.) As late as 1911 the *Encyclopedia Britannica* dismissed Vasari's allegation on grounds like these.

By this time, however, Piero's mathematical works had already been discovered. The first volume was found by G. Pitarelli in 1903 in the Vatican Library. It proved to be a work on the Platonic solids, ceremoniously dedicated to Duke Guidobaldo, the son of Federigo, to be shelved with *De Prospectiva Pingendi* in the great library at Urbino. Any doubt that it was Piero's vanished when the marginal corrections were found by G. Mancini to be in Piero's own hand.[2] Another mathematical work by Piero came to light soon after, known now as the *Trattato d'Abaco*, and it survives in several manuscript copies. Just as Vasari had said, both books are incorporated into the printed books of Luca Pacioli without attribution.

So it seems that something had happened to mathematics by about 1500. Mathematics was now worth stealing. It wasn't clear what mathematics was, but that meant you could invent meanings for it. With the invention of printing, erudite speculations about this most tantalizing of subjects could be going hand to hand, with your name on the cover.

The Abacus Tradition

Both Piero della Francesca and Luca Pacioli came out of the abacus tradition. Contrary to what one might think, the abacus schools did not teach students to use the abacus, but rather to use a system of computation on paper that *replaced* the abacus, essentially the same system of doing sums on paper that we still use today. It was introduced into Italy around 1200 by Leonardo Pisano, later called Fibonacci. As a boy, Leonardo had lived in North Africa and learned the mathematics of the Arabs, as he tells us himself in his *Liber Abbaci:*

> I joined my father after his assignment by his homeland Pisa as an officer in the customhouse located at Bugia [Algeria] for the Pisan merchants who were often there. He had me marvelously instructed in the Arabic-Hindu numerals and calculation. I enjoyed so much the instruction that I later continued to study mathematics while on business trips to Egypt, Syria, Greece, Sicily, and Provence and there enjoyed discussions and disputations with the scholars of those places. Returning to Pisa I composed this book of fifteen chapters which comprises what I feel is the best of the Hindu, Arabic, and Greek methods. I have included proofs to further the understanding of the reader and the Italian people. If by chance I have omitted anything more or less proper or necessary, I beg forgiveness, since there is no one who is without fault and circumspect in all matters.[3]

Leonardo's conscientious and pathbreaking work stood the test of time very well. For 300 years and more, practical mathematics in Italy was the study of books modeled on his. There is remarkably little change over the centuries. Even nineteenth-century American arithmetic textbooks seem to be in the abacus tradition of Leonardo Pisano. All these books introduce the numerals 0,1,2, . . ., and explain place notation for representing numbers. They continue with the operations of arithmetic, which can now be done on paper, and finish with practical problems of buying, selling, and lending money at interest. The abacus schools produced a mercantile class well drilled in practical mathematics, the foundation of Italy's wealth. The trading and financial economy that made the Italian Renaissance possible owed Leonardo Pisano an incalculable debt. As an apocryphal story attests, a German merchant counseled his son that if he was content to learn just addition and subtraction, a German university would be good enough, but that if he wanted to learn multiplication and division, he would have to go to Italy.[4]

Leonardo wrote other books as well, and one of these, the *Practica geometriae*, also contributed to the abacus curriculum. Students learned to find areas and volumes of simple shapes, given their dimensions.

Piero della Francesca wrote a *Trattato d'Abaco*, as we have said, but within the genre of abacus texts it is a very peculiar book. It is not so clear who the audience is—certainly not little boys. It begins, very tersely, "Having been asked to write a few things about mathematics useful to merchants, by one whose wishes are my commands, not through presumption but to obey him, with the help of God, treating some commercial problems like barter, contracts, and partnerships, beginning with the Rule of Three and, God willing, some algebra: and first how to multiply and divide fractions." By the third sentence of the book, Piero is absorbed in the rule for multiplying fractions. The culmination of the usual abacus course is the Rule of Three, but Piero gets to it by page five. The problem he starts with is not chosen as a simple illustration, and it seems needlessly complicated. It asks, if 7 loaves of bread cost 9 pounds, how much do 5 loaves cost? Answer: 6 pounds, 8 soldi, and $6\frac{6}{7}$ denari. Many such problems follow, all worked out in detail, in the abacus tradition, but at an unusual level of arithmetical complexity.

The *Trattato d'Abaco*, like many other abacus texts, has what looks like a section on practical geometry, but it is at a very high level and includes things not found in any other abacus book, like an algorithm for finding the volume of a general tetrahedron given its six edges. This problem was apparently forgotten and not solved again until 1841, by Arthur Cayley. In 1852 the American mathematician J. J. Sylvester wrote, "Query, Is not this expression for the volume of a pyramid in terms of its sides to be found in some previous writer? It can hardly have escaped inquiry."[5]

By page thirty or so of the *Trattato d'Abaco*, Piero is posing problems of algebra that no one of the time knew how to solve, such as cubic equations. It is not clear that there was any audience for this, but since the book survives in several copies, someone had found it worth copying over. Piero, alas, makes no significant progress on these algebraic problems, even though he proposes solutions that work in special cases. The attempt, though, is a surprise. Piero was working at the forefront not only of geometry but also of algebra.

The appearance of cutting-edge mathematical problems in what claims to be an abacus text lets us know that there was no established place for new mathematics in the fifteenth century. The culture simply had not provided for original work in mathematics. The curriculum of the abacus school was frozen where it had been since the year 1200, and philosophical mathematics was frozen where it had been since the ancient times of Nichomachus and Boethius. The theory of transformation that Piero called degradation, and that would later be called projective geometry, was hidden, as we have seen, in a manuscript handbook for painters. And as we will see in the next section, Piero seems to have considered his final mathematical book, the one on the Platonic solids, to be essentially a work of art. He presented it to the Duke of Urbino, in a unique manuscript copy, translated into Latin for *gravitas*, as he might have presented a painting.

The Book on the Five Regular Solids

Piero's little book on the Platonic solids, *Libellus* for short, formulates and solves a large number of problems having to do with two kinds of geometry. Piero had learned abacus school geometry as a child, but later he learned Euclidean geometry. The two are really quite different. Where Euclidean geometry proves theorems and contains no arithmetic at all, abacus school geometry, so-called practical geometry, is about finding areas and volumes, or sides of triangles, by arithmetic. Where Euclidean geometry might construct a regular pentagon, practical geometry provides a formula for its area given its side.

Abacus school geometry covered only the simplest figures of Euclidean geometry and did not go as far as the Platonic solids. In the *Libellus* Piero extended arithmetic to the Platonic solids, showing, for example, how to find such things as the volume of an icosahedron. That question occurs as Problem 37 of Book II of *Libellus*, where Piero shows that an icosahedron of volume 400 has the side $x = \left(806400 - \sqrt{597196800000}\right)^{1/6}$. This is a bizarre number. It turns out to be more than 5 and less than 6, but who could tell that by looking at it? There is no practical motivation for computing this number. Why is Piero doing it?

Whatever the reason, Piero put together 139 such problems and presented them to Duke Guidobaldo of Urbino, son of the glorious Federigo, in a book to be shelved with *De Prospectiva Pingendi* in the Urbino library. The dedication to Guidobaldo begins in a rather formulaic way, linking Federigo and Guidobaldo to the great patrons of the classical past, pointing out that it is the glory of a patron that makes a work of art truly great, but it then becomes unexpectedly personal and touching. This book, says Piero, is the work of my old age, the last fruit of an exhausted acreage, from which your father had more abundantly (a reference to his many paintings for Federigo). Let this book, just a little something that I did to keep my mind from going torpid, be my memento to you, in token of my perpetual service.

The *Libellus* is, in effect, a work of art. Not only does Piero present *Libellus* to Guidobaldo as he had presented paintings to Federigo, but he even cites the age-old relationship between artist and patron, although the art is now mathematics. In fact, since the book was carefully written out in a nice hand, and included beautiful line drawings, it *was* a work of art if you forget its contents and just think of it as an object. The book can sit modestly on a shelf "in a corner," says Piero, because "it is not the custom to place on an opulent table mere apples brought in by a rustic peasant." He does, however, suggest that people might come to see the book, just as they might come to see any other work of art. "It might be pleasing in its novelty," he says, "the geometry of Euclid, but translated for arithmeticians." And Luca Pacioli, for one, did come to see it.

What is this little book for? It is odd, as so much of Renaissance mathematics is odd, and Piero presents it to Guidobaldo without much explanation. It is arguably some of the first original mathematics in Europe after Fibonacci, but except for the sharp-eyed Pacioli, no one seems to have noticed it. The book itself was virtually forgotten. Vasari suggests that *Libellus* should have contributed to Piero's fame and honor, but fame in mathematics was hardly thought of in Piero's time. That whole concept owes a lot to Pacioli, in fact, who did make himself famous with mathematics, but only after Piero's death.

The idea of extending practical geometry to more complicated geometrical figures makes a kind of pure mathematical sense. Indeed, Leonardo Pisano himself had completed a few such problems involving the Platonic solids 250 years earlier, in his original *Practica Geometriae*. They did not get incorporated into the abacus school curriculum because they had no practical application (and they were too hard). But for anyone who could accept this challenge, it was obvious that problems about the Platonic solids were worthy problems in their own right, having something to do with fundamental, natural, mysterious objects. You could work on them for more than just your own moral elevation, or even your own fame. You could work on them because they were mysteriously important. This is the spirit that drives research in pure mathematics. The simplest explanation of *Libellus* is that it is pure mathematics, and that the odd presentation to Guidobaldo is something of an afterthought, with no expectations.

The court of Guidobaldo at almost precisely the time of Piero's presentation of *Libellus* is delightfully recreated in Baldassare Castiglione's *The Courtier*. The book recreates the witty conversations of an aristocratic elite at play, as they tease and challenge each other, all in the name of imagining the perfect courtier and the perfect court lady. These paragons should not lack for any good quality, although in what proportions they should have them is a matter for refined judgment to consider. The courtier should play chess, for example, but not too well, lest it be thought that he considered it important. The boy Raphael, already revealing his excellence as an artist, is mentioned as a court favorite, and so far as art goes, the courtier should know how to appreciate pictures. This, however, is as close as the courtier ever gets to mathematics, which is never explicitly mentioned. Mathematics does not occur to anyone as an accomplishment that an educated person should have, and this in one of the only places where original mathematical work had actually been going on. Many of those in attendance had probably seen old Piero shuffling around the library, but he apparently made no impression on them at all. Pietro Bembo, a prominent personality in *The Courtier*, does mention some years later that his father had gone to a lecture by Luca Pacioli, but it was perhaps just out of curiosity to see a novelty. New discoveries and the medium of print were just beginning to make mathematics visible.

Piero's own work was almost completely invisible at the time, but it was quite daring. All of Piero's mathematics pushes on the boundaries of what was known to him. The purpose of the *Libellus* may have been unclear, but after all, it is in the nature of research that you do not know for sure where it will go. Many of the problems in *Libellus* follow the same plan and might seem repetitive and tedious, but then in Book III, Problem 9, Piero's patience pays off. Just by following his routine methods, Piero finds a previously unknown and very pretty theorem of Euclidean geometry, relating the diameter of the dodecahedron to its pentagonal face. The method of proof is by arithmetic, and is therefore completely unlike Euclid's proofs. This is a significant development from the purely mathematical point of view, as the two branches of quadrivium mathematics, geometry and arithmetic, are here beginning to mix and blur. Piero computes the diameter of the dodecahedron numerically (edge to opposite edge), and then notices that he gets the same number by adding the side and chord of the pentagonal face. The result follows.

The dodecahedron theorem comes out of patient workmanship and close attention. But another result, Book IV, Problem 10, comes seemingly out of nowhere, and it is stated and solved without any motivation. Piero finds the volume of the cross-vault. The cross-vault is the chamber formed by two cylinders intersecting at right angles, an architectural feature of cathedrals and palaces occurring where two round, arched corridors intersect. Piero solves this problem by reference to a result of Archimedes, so he must have worked it out in the Urbino library. The way he uses Archimedes isn't obvious, and Archimedes himself does not address this problem. In Book IV, Problem 11, Piero finds the surface area of the interior curved walls of the cross-vault, another surprisingly sophisticated result. These are problems that challenge today's students, who can use calculus, the natural tool for the job—a tool that Piero did not have.

In a peculiar twist, it turns out that Archimedes had addressed the problem of the cross-vault after all, although Piero couldn't have known this. That fact came to light only in 1906 when Theodore Heiberg discovered a tenth-century palimpsest in Constantinople containing both known and previously unknown writings of Archimedes. It is truly astonishing that Piero seems to have read Archimedes's mind here, as there is no possibility that this work of Archimedes was known in the fifteenth century.

Beyond the results of this or that individual problem, though, there is a larger question behind what Piero is doing. His computations typically produce square roots, and even higher roots, like the side of the icosahedron, which involves a square root and a sixth root. Since the time of Pythagoras it had been known that such numbers are typically not simple fractions. The diagonal of the unit square, to take the simplest case, is the number $\sqrt{2}$, a square root. This is approximately $7/5$, but not quite that. A better approximation is $17/12$, but it is not quite that either. In fact there is no fraction of this kind that represents $\sqrt{2}$. This square root is said to be *irrational,* and the diagonal and the side of the square are said to be *incommensurable.* This ancient complication in arithmetic seemed to say that familiar numbers couldn't always represent the quantities of geometry. It meant that geometry and arithmetic were quite different.

If you include the irrational square roots, though, together with the more familiar numbers, you get a system of numbers containing square roots and higher roots that does seem to be adequate to geometry, at least in simple cases. Perhaps one of Piero's motivations was to push out into the sea of more challenging geometry problems to see what would happen with these irrational numbers. Would they continue to make sense?

A good example of Piero's investigation of irrational numbers is *Libellus* Book III, Problem 1. The problem is to investigate an octahedron inscribed in a tetrahedron with side 12. That sounds complicated, but it turns out that if the tetrahedron has side 12, then the octahedron has side 6—Piero's picture makes it obvious. Rather than simply noting this fact and going on, Piero, for some reason, provides for this problem a bizarre solution, the most complicated one in the entire book. He drops altitudes in the faces and through the interior, including one in a very asymmetric position. He notes nonobvious similar triangles, and at one point must recognize $\sqrt{108} - \sqrt{12} = \sqrt{48}$. In the end he finds the side of the octahedron as $\sqrt{36}$. This is, of course, correct, if you agree that $\sqrt{36} = 6$. This particular square root is not irrational after all, although it was found by way of a lot of other square roots that *were* irrational. When all these dubious irrational numbers have been subjected to the operations of arithmetic, the answer comes out right. The arithmetic of irrational numbers seems to work, and it agrees with geometry.

From a modern point of view, this investigation of irrational numbers is unsophisticated. Even if Problem I in Book III is suggestive, it is just one problem, and it doesn't prove anything. Mathematicians know, though, that this is how you explore the unknown—you find an example where you think you know what should happen, and then you see what happens with the mathematics. Piero's active pushing and pulling on the objects of classical mathematics was a sensible way to begin to find out what they really were. In the mathematics of Piero, a modern mathematician sees the glimmering of something that makes sense.

The same cannot be said of the mathematics of his countryman, Luca Pacioli. With a little insight into Pacioli's character, though, we can begin to imagine how covetously Pacioli must have studied *Libellus*. The irrational numbers fascinated him, the Platonic solids fascinated him, and the transcendental importance of these things became his stock in trade.

Luca Pacioli and the Golden Ratio

According to Vasari, Luca Pacioli was Piero della Francesca's student, but that can't be true. Piero was already working on artistic commissions in other Italian cities when Pacioli was born in San Sepolcro, around 1445. Pacioli himself says that he began his study of higher mathematics in 1464 with Domenico Bragadino in Venice. Bragadino gave public lectures on mathematics, and soon Pacioli was doing the same thing. He became a Franciscan friar sometime before 1477, and it is known that he had access to the works of Euclid in a volume that he borrowed from the convent of Saint Francis in Perugia, returning the Euclid in 1480.[6] Pacioli had a reputation as a powerful preacher, and more than just the Word, he preached mathematics.

Part huckster, part charlatan, part true believer, this self-promoter, entrepreneur, and opportunist offers us a window onto Renaissance mathematics in all its idiosyncrasy. He showed by his example that it was possible to make the leap into the privileged class by promoting mathematics. Generations of mathematicians after him flourished outside the universities as independent agents, living by their lectures, their tutoring, and their books. (Over a hundred years later Galileo would encounter Euclidean geometry not at the university, but from an enterprising lecturer with a position at court, Ostilio Ricci.) Pacioli understood the power of print, as Piero did not, and his first printed book, the *Summa Arithmetica,* seems to contain essentially everything that he could get his hands on, in no particular order, including Piero's *Trattato d'Abaco,* but without any indication that much of the contents of the *Summa* had been written by other people. His fawning dedications succeeded in winning him highly placed protectors who were able to intervene for him in the crises of his life.

The humanist project of recovering the classical past did not ignore mathematics. Essentially all the classical mathematical works we know today were known then, but the significance of these works was obscure, even while it was widely felt that they might be crucially important. Pacioli made this mystery of mathematics, which he felt powerfully himself, the subject of public lectures and private lessons, and he did much to heighten popular interest in mathematics, though he increased rather than decreased its mystery. He had good reasons for wanting to heighten its mystery, because it added to the drama of his lectures, and he could not in any case have dispelled it. His own understanding, despite his sincere passion and ambition, was very shallow.

He is a particularly energetic representative of a literate population that felt the pull of mathematics, bought mathematical books, went to mathematical lectures, and were intensely interested in mathematics, but whose contribution was their enthusiasm and their support of a continuing mathematical culture rather than any addition to mathematics itself.

One of his mathematical sermons, delivered August 11, 1508, in the Church of St. Bartholomew in Venice, is preserved verbatim because he included it in his printed edition of Euclid the following year as an introduction to Book V on Eudoxian proportion.[7] "Most difficult is proportion: this it is which alone gives access to the innermost nature of the most high and indivisible Trinity, and it should be investigated most zealously by the doctors of sacred theology."[8] Knowledge of proportion has been sought, he says, by philosophers of the divine, by metaphysicians, and by natural philosophers, Socrates, Plato, and Aristotle. He enumerates the arts and their famous practitioners, medicine, astronomy, cosmography, architecture (Archimedes is placed among the architects for some reason, probably because Pacioli knows him only through Vitruvius), painting, sculpture, music, poetry, oratory, and justice, reminding his listeners on the matter of justice that "the most righteous judge of the living and the dead will one day reward each human creature *in proportion* to his merits and demerits, as sacred scripture clearly tells us."[9] It is ironic that on just the point that Pacioli felt was most important, the translation he was reprinting was incomprehensible: it garbled the crucial Definition 5.[10] On the other hand, the complete mystery of that passage might have been what Pacioli found most attractive about it.

An example of Pacioli's popular mathematical mystery culture that survives to this day is the cult of the golden rectangle, or the golden section, or the golden ratio. These are all names for the same idea. One hears that the golden rectangle, with sides in the golden ratio ϕ, as it is called, is the most aesthetically perfect of rectangles. Renaissance painters are supposed to have used the golden rectangle in planning their pictures, the Parthenon is supposed to be ordered by the golden rectangle, the Great Pyramid of Egypt encodes the golden ratio ϕ, and so on.

The popularization of ϕ and its importance goes back to Luca Pacioli's book *De Divina Proportione,* the Divine Proportion. This book is apparently responsible for the enthusiasm that continues, in some manner, at least, to this day. Pacioli never suggested that Renaissance painters used the golden ratio ϕ, because he would have been contradicted immediately by the painters themselves. He never suggested that the golden rectangle is aesthetically perfect, or that the Parthenon is a golden rectangle, or any of the other versions of the idea that are familiar today, but he did start the process that was taken up by perhaps unwitting disciples throughout the centuries who may never have heard of Pacioli. Anyone who can launch an idea so enduring, even if it is self-serving invention, deserves our respect.

The golden ratio arises in connection with regular pentagons. The occult symbol of the pentagram, a regular five-pointed star inscribed in a circle, shows the golden ratio nicely. Each side of the star cuts two other sides of the star into a long piece and a short piece. These pieces are in golden ratio ϕ.

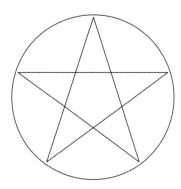

The sides of the star in the pentagram cut each other in golden ratio.

The line segment L is divided in golden ratio, the golden section.

Another way to obtain the golden ratio is to divide a line of length L into two pieces a and b such that a is the same fraction of b that b is of L. The golden ratio is then the ratio of b to a, or equivalently of L to b, and turns out to be numerically about $1.618\ldots$ It is actually the irrational number $\left(1+\sqrt{5}\right)2$, and thus cannot be written down exactly as a simple fraction.

Euclid's construction of the regular pentagon begins by dividing a line in golden ratio (which Euclid calls simply "mean ratio"), although in Euclid the operation of dividing a line in mean ratio has no special significance. In the Neoplatonic world of the fifteenth century, though, this operation was pregnant with potential meaning. The regular pentagon was, after all, the building block of the dodecahedron, which, according to Plato, was the atom of celestial matter. The secrets of ϕ might be cosmically significant.

Pacioli milked the significance of ϕ in *De Divina Proportione*. After a sonnet in praise of the Platonic solids, there is a little poem in which the Platonic solids themselves answer back: "You who strive to investigate the reasons of things, learn! Study us! This way is open to everyone!" Pacioli justifies his title, calling ϕ the divine proportion, because ϕ relates three things together, namely a, b, and L as shown in the figure, and for this purpose three quantities are necessary, neither more nor less. Thus ϕ is an expression of the Trinity: Father, Son, and Holy Spirit. What is more, ϕ is eternally what it is and will never change, like God Himself. And what is still more, our proportion ϕ cannot be defined as a fraction because it is irrational, just as God cannot be defined in words.

Thirteen marvelous properties of ϕ follow, in thirteen little chapters, concluding with the observation that there wouldn't be paper and ink enough to say everything that could be said. But since Jesus Christ, together with His twelve disciples, made thirteen, then the number thirteen, Pacioli says, is good enough for him. The thirteen marvelous properties of ϕ turn out to be trivial algebraic properties of the number ϕ, but they are all stated breathlessly and portentously, as in "its fourth inexpressible property . . . its seventh incredible property . . . its ninth property, exceeding all the preceding ones . . . its twelfth nearly inconceivable property . . . its thirteenth most precious property."

Pacioli next describes the five Platonic solids and how they nest into each other. He reminds the reader that the five Platonic solids are the atoms of the five elements, according to Plato, and even says, in language that must have raised eyebrows, that if a sixth one could be made, "the Creator would be diminished in his creation, and we would have to judge Him lacking in foresight, in not recognizing from the beginning what was necessary and appropriate."

It is hard to know how seriously to take this inflated language, but Cardinal Pietro Soderini, to whom the book was ultimately dedicated, seems to have felt highly honored. The religious language sounds borderline blasphemous, but apparently no one objected to it. Mathematics and the divine were still very easily and naturally related in the early sixteenth century, it seems. And as we have noted already, the mystical cult of ϕ continues to this day.

Pacioli has a great deal more to say about the importance of the Platonic solids for the enlightened ruler. "By this infallible method, Your Highness can have fun with stonemasons, as we have done ourselves." Pacioli tells a surprisingly nasty little story about himself in Rome, visiting a certain Duke Girolamo, who was building a palace. He and the duke agreed to expose the ignorance of a stonemason, asking him to make a regular polyhedron for the capital of a column, stipulating that it should have twenty faces, but that they should not be triangular. Could the workman do it? Of course I can do it, he said scornfully. But he had agreed to an impossible task, because it is known from Euclid that the only regular polyhedron with twenty faces is the icosahedron, and it has triangular faces. And so, after many days and many spoiled marble blocks, the workman had to admit that he could not do it. He knew that it was the friar who had caused his embarrassment, but instead of being resentful, he begged Pacioli to explain the matter to him, and Pacioli was glad to teach him. And when he was done, this man was so grateful that he gave Pacioli a nice coat.

Perhaps the peculiarity of this passage should just stand on its own as a piece of Renaissance mathematics in all its oddity, but one could also read a little more into it. Pacioli was born into the middle class, but by his own energetic efforts he had lifted himself up. He frequently lectured in universities, although he never stayed long at any one.[11] He revels in mentioning the names of famous people he has met. In this story he drops the name of Duke Girolamo, but there is an unmistakable nervousness in the account. Surely it isn't the duke who wants to expose the ignorance of the stonemason. The stonemason is depicted as a buffoon, though an educable one, but Pacioli might have seen such people in reality as a threat. A stonemason is a low-level artist and sculptor, and Pacioli knew by Piero's example that such people could actually know a lot, more than Pacioli himself. Their mathematical insight came not from books but from experience, something that Pacioli had no conception of, to judge by his writing. The importance of the arts for mathematics was perhaps beginning to be dimly apparent even then. In this case Pacioli gets high marks for perception, even if it was the perception of a threat.

Luca Pacioli means to depict himself as a wise and conscientious teacher and advisor, and *De Divina Proportione* is written in the form of an illuminating and helpful, or perhaps even indispensable, contribution for the benefit of the illustrious court of Duke Ludovico Sforza of Milan. Pacioli came to Milan in 1498, and from his opening chapter, where he describes his own arrival, we can easily picture how he must have ingratiated himself with the Milanese court. He names all the courtiers and praises their admirable qualities, but he dwells longer on one name than on any other, a man with whom he had struck a fateful, and on his side perhaps uneasy, friendship: Leonardo da Vinci.

Leonardo da Vinci

"Sometimes, in supernatural fashion, beauty, grace, and talent are united so far beyond measure in one single person, that wherever he turns his attention, his every action is so divine, that, surpassing all other men, it reveals itself clearly as a gift of God. This was seen by all men in Leonardo da Vinci." With these words Giorgio Vasari begins the most extensive of his many biographical sketches of the great artists of his time. Besides genius in drawing and painting that has never been equaled—Vasari says of the Mona Lisa that "it was painted in such a manner as to make every valiant artist, without exception, tremble and lose heart"—Leonardo possessed a personal charisma that was acknowledged by everyone, and that never left him. Even in the twilight of his life, in the comfortable retirement offered him by King Francis I of France, he was the favorite of the king, who visited him almost daily.[12] According to Benvenuto Cellini, not usually one to take a back seat, the king "believed no other man had been born who knew as much as Leonardo, both in sculpture, painting, and architecture, so that he was a very great philosopher."

Leonardo's reputation as a philosopher did not reflect his learning, as he had had almost no formal education. Even at the level of the abacus school he may not have had much instruction. According to Vasari, "in arithmetic, during the few months that he studied it, he made so much progress that, by continually raising doubts and difficulties with his teacher, he would very often bewilder him." Leonardo was a lively and intelligent pupil, we can be sure, but the instruction apparently lasted only a few months.

All his life Leonardo was defensive about his lack of education, in the classics especially, calling himself "an unlettered man." His characteristic rebuttal to his (imaginary?) detractors was, "Though I have no power to quote from authors as they have, I shall rely on a far greater and more worthy thing—on experience, the instructress of their masters. They strut about puffed up and pompous, decked out and adorned not by their own labors, but by those of others, and they will not even allow me my own." For Leonardo the immediate experience of the senses, and especially the sense of sight, was the surest basis for knowledge, "for if we are doubtful about the certainty of things that pass through the senses, how much more would we question the many things against which these senses rebel, such as the nature of God and the soul and the like?" Translating from the senses, Leonardo used drawing in the way that Dante had used poetry, as a tool to discover the truth of things, whether fleeting, hidden, or obscure, from human gesture, to anatomy, to the turbulent flow of water.

In his thirst for discovery Leonardo gradually became convinced of the importance of mathematics, for the certainty of its propositions above all else. He even wrote, perhaps wistfully, imitating the inscription over Plato's Academy, "Let no man who is not a mathematician read the elements of my work." In truth, there is no real mathematics in Leonardo's work, but his yearning for that work to have the certainty and the solidity of mathematics is palpable.

Leonardo's early artistic training and career were in Florence, but in 1482, when he was thirty, Lorenzo the Magnificent sent him on a delegation to Duke Ludovico Sforza in Milan—as a musician! He performed on a lyre of his own design to great applause, and stayed on in Milan as one of the brightest luminaries of that luminous court. Paintings made there include the portrait of a lady with an ermine (Cecilia Gallerani, the duke's mistress) and of course *The Last Supper*.

Leonardo was not primarily a painter in Milan, though. He worked very slowly, even if he was seriously engaged with a painting, and he took long absences from work that he was supposed to be completing, having many other things on his mind. The duke, in turn, had many other things to occupy Leonardo, regarding him as an all-purpose engineer and craftsman. Leonardo had described himself this way in a letter to the duke, offering especially his secret plans for war machines: "1. I have plans of bridges, very light and strong . . . 2. When a place is besieged, I know how to remove the water from the trenches . . . 3 . . . I have plans for destroying every fortress or other stronghold even if it were founded on rock. 4. I have also plans of mortars most convenient and easy to carry with which to hurl small stones in the manner almost of a storm . . ." It has been argued recently that Leonardo is actually responsible, indirectly at least, for the wheelock, or flintlock, the mechanism of a musket.[13] This idea looks suspiciously like his fourth proposal in the letter to Ludovico. In any case it seems clear that Leonardo continued to work on such things after his arrival in Milan and that they weren't just fantasies, even if he never saw this or any other of his inventions come to fruition.

Only after a long list of such ideas does he say, "I can carry out sculpture in marble, bronze or clay, and also I can do in painting whatever it is possible to do, as well as any other man, be he who he may."[14] One famous project that Leonardo undertook in Milan combined engineering and artistic challenges, the erection of an enormous equestrian monument to Duke Ludovico's father.

When Luca Pacioli arrived in Milan in 1498, Leonardo had been there for over fifteen years. It is easy to imagine the eagerness with which Pacioli cultivated the famous and beautiful Leonardo, whose *Last Supper* had just been completed (as much as it would ever be completed), open to view in an unspoiled perfection that we can only dream about now. But it is also not hard to understand how Leonardo might have gravitated toward Pacioli. By his own account this friar, Maestro Luca, as he would have called him, had what Leonardo craved, a deep understanding of the mathematics that underlay appearances. They became friends. Within weeks Pacioli had enlisted the incomparable hand of Leonardo da Vinci to illustrate *De Divina Proportione*. In the book Pacioli drops Leonardo's name at every opportunity, and in one of the margins he draws a line *ab* that only needs to be multiplied by $37\frac{1}{5}$ to show the height of Leonardo's planned equestrian monument.

It is painful to think how vulnerable Leonardo might have been to the seductions of Pacioli's geometry. In the end, Leonardo produced sixty careful drawings of the Platonic solids and other polyhedra, representing physical models of these figures fashioned as hollow frameworks, so that you could see through them to the back. Leonardo was at the height of his powers—the *Mona Lisa* was still to come—but here he seems to be doing something utterly mechanical, and almost embarrassing. He must have been glad to do it, though. That he finished the job at all is proof of this—he never did anything against his will. And Pacioli cannot be exaggerating the extent of Leonardo's contribution, because all of this was published in Leonardo's lifetime, and he could easily have disclaimed it. Increasingly after this time Leonardo's notebooks are filled with geometrical doodles of no apparent mathematical significance, but which seem to have significance for him. Leonardo became increasingly preoccupied with mathematics, sometimes to the extent of losing interest in everything else, even as his progress in understanding mathematics is impossible to discern. A notebook entry apparently from 1506 says "learn the multiplication of square roots from Maestro Luca."

The glory of the Milanese court came to an abrupt and catastrophic end less than two years after Luca Pacioli's arrival. The French were at the gates, and the metal destined for the equestrian monument went for cannon instead. In October 1499, a French army entered Milan, and Duke Ludovico's patronage was at an end. He was sent as a prisoner to France. French gunners used Leonardo's clay model of the great horse for target practice. Leonardo da Vinci and Luca Pacioli traveled together to Venice and apparently remained in touch long after that. Leonardo served as an engineer with Cesare Borgia (to whom Niccolò Machiavelli addressed *The Prince*) and was present when Cesare took Urbino in 1502 and sacked the ducal library. Luca Pacioli kept Leonardo's drawings and prepared plates for the printed version of *De Divina Proportione*, published finally in Venice in 1509. All the vanished pomp of the Milanese court is depicted there as if it still existed, and Pacioli extols the virtues and ineffable properties of his mathematics as if nothing unpleasant had happened to Duke Ludovico, invoking without irony the great example of Archimedes and mathematics' finest hour at the siege of Syracuse.

Music

Galileo's father Vincenzo Galilei knew very well the storied power of Greek music. He knew, for example, how Pythagoras had calmed "a Taoromenian lad who, after feasting by night, intended to burn the vestibule of the house of his mistress, on seeing her coming out of the house of his rival. The lad had been inflamed by a Phrygian song, but Pythagoras, happening to meet this Phrygian piper at an unseasonable time of night, persuaded him to change his Phrygian song for a spondaic one. Through this the fury of the lad was immediately soothed, and he returned home in an orderly manner, although but a little while before he had stupidly insulted Pythagoras on meeting him, and would bear no admonition, and could not be restrained."[1] Pythagoras could work such effects because he understood that "the primal understanding is that of music's melodies and rhythms. From these he obtained remedies of human manners and passions, and restored the pristine harmony of the faculties of the soul."[2]

Why was the music of Pythagoras's time so effective, and yet the music of Galilei's own time, in his mind at least, so empty and frivolous? It became the ruling passion of Vincenzo Galilei's life to understand this question. Not a man of much means, Galilei was fortunate to have the support and friendship of a far more powerful patron, Giovanni Bardi, who shared his enthusiasm. Bardi was the sort of person to be put in charge of the music for a Medici wedding, the kind of extravagance that is talked about not just for a season but for centuries. That is to say, Bardi was the sort of person to make things happen, and he made it possible for Galilei to research the question of Greek music as thoroughly as it was possible to do. He put Galilei in touch with a Florentine humanist, Girolamo Mei, in Rome, who had thought deeply about this question. Bardi vouched for him when Mei wanted to be sure that Galilei really was a gentleman of good character. He financed research trips for Galilei and helped him procure musical instruments from distant places. Galilei in his turn showed great imagination in his use of ethnographic evidence, and records preserved in visual form, like ancient bas reliefs depicting musicians performing. The progress of this research was a frequent topic of discussion at the Florentine Camerata, an intellectual musical society that met at Bardi's house.

Every member of the Camerata would have known the special place of music in Plato's *Republic*, and in the education of young children. Even before a child is old enough to reason,

> rhythm and harmony [should] sink deep into the recesses of the soul and take the strongest hold there, bringing that grace of body and mind which is only to be found in one who is brought up in the right way. Moreover, a proper training of this kind makes a man quick to perceive any defect or ugliness in art or in nature. Such deformity will rightly disgust him. Approving all that is lovely, he will welcome it home with joy into his soul and, nourished thereby, grow into a man of a noble spirit. All that is ugly and disgraceful he will rightly condemn and abhor while he is still too young to understand the reason; and when reason comes he will greet her as a friend with whom his education has made him long familiar.[3]

Galilei's own fascination with the moral force of music echoes, in a contemporary way, that of Plato, and tells us, no doubt, something about how his son Galileo was educated, not just in music, but in general, to an intellectual appreciation of the beauty of the world.

The way Greek music attained its effects was not entirely mysterious. There was actually a surprising amount of technical information extant about it. A discussion between Socrates and Glaucon about which musical modes would be suitable for the ideal republic is quite specific. We will want no dirges and laments, they say, so that eliminates Mixed Lydian and Hyperlydian modes. No drunkenness or effeminacy, so forget Ionian and modes of that kind. "That leaves Dorian and Phrygian," said Glaucon.

> "I am not an expert in the modes," said Socrates, "but leave me one which will fittingly represent the tones and accents of a brave man in warlike action . . . who, in the hour of defeat or when facing wounds and death, will meet every blow of fortune with steadfast endurance. We shall need another to express peaceful action under no stress of hard necessity; as when a man is using persuasion or entreaty, praying to the gods or instructing and admonishing his neighbor, or again submitting himself to the instruction and persuasion of others; a man who is not overbearing when any such action has proved successful, but behaves always with wise restraint and is content with the outcome. These two modes you must leave."[4]

"The modes you want," said Glaucon, "are just the two I mentioned." One naturally wonders, then, what was Dorian mode? What was Phrygian mode? What did they sound like?

Here too, there was a surprising amount of technical information available, largely because of the association of the musical intervals with arithmetic ratios, the old discovery of Pythagoras. The association between music and astronomy tended to make the subject of music even more scientific, and more likely to be treated mathematically and objectively in a form that could be reconstructed, at least in theory. And that is what Galilei strove to do.

7 The Orphic Mystery

The music of the distant past might seem to be lost forever, but in the case of Greek music, the surviving evidence is tantalizingly abundant and informative. The intervals of Greek scales and modes are described mathematically in several sources, and there is a system for notating melodies that survives in tables. There are even a few surviving specimens of Greek music written in this notation. Vincenzo Galilei published four short ancient Greek hymns, ascribed to the second-century lyrist Mesomedes, in his *Dialogue on Ancient and Modern Music* from manuscript copies provided to him by Girolamo Mei.[1] Other fragments of ancient Greek music continue to turn up on papyri, and in 1893 during excavations at Delphi two hymns were found engraved in stone.[2]

In spite of all this, no one claims to know with any confidence what Greek music sounded like. Even if we did know that, we still would not hear it with the cultural resonances that it had for its original audience. A living musical tradition cannot be captured in notation. Try to imagine what a Shostakovich string quartet might sound like in a future reconstruction by a civilization that had some idea what our notation meant and what violins looked like, and compare that exercise with the bravura of a modern performance by a really good quartet at the height of its success before an adoring contemporary audience.

And yet—isn't there the possibility of a direct link to the past? As late as 400 c.e. the Christian Neoplatonist Synesius of Cyrene, many of whose letters survive, speaks of singing the hymns of Mesomedes to the lyre.[3] And barely 200 years later, Pope Gregory I codified the plainchant music of the church, what we now know as Gregorian chant. Later systematizations of Gregorian chant gave to the chant modes the old Greek names: Dorian, Lydian, and so on. Might Gregorian chant actually be Greek?

There is even a musical instrument with a continuous history of use through this period, the pipe organ. Far too complex to have been a creation of the Middle Ages, the pipe organ is a Hellenistic invention of the hydraulic engineer Ctesibius, according to Vitruvius. It was popular in Rome and Byzantium, and it never disappeared. Its scales, even if they evolved somewhat over the centuries, must be essentially ancient Greek scales.

The Problem of Tuning

All questions about tuning and scales must start from Pythagoras's original observation that there are certain intervals that sound peculiarly consonant and sonorous, namely the octave (2 : 1), the fifth (3 : 2), and the fourth (4 : 3). The ratios indicated are the ones assigned by Pythagoras on the basis of experiments with stretched strings. They correspond to the lengths of strings at fixed tension, and also, we now know, to the frequency of the notes sounded by the strings. The frequency goes up by a factor 2 when the length is shortened by a factor 2, so that the note goes up an octave. Likewise, the frequency goes up by a factor $3/2$ when the length is shortened by a factor $3/2$ (in other words, it is only $2/3$ the length it was before)—the note goes up by a musical fifth. The frequency goes up by a factor $4/3$ when the string is shortened by a factor $4/3$ (when it is only $3/4$ the length it was before), and so the note goes up by a musical fourth.

Galileo, in his last book *Two New Sciences*, is still interested in this matter of pitches and ratios. He offers a new observation in support of the idea that what we hear as an octave is really a factor 2 in frequency. He describes using a chisel on a metal plate, causing the metal to "sing," while at the same time dust on the plate collected in patterns showing the nodes of the vibrations. Suddenly the chisel slipped and excited the note an octave higher, while at the same time the dust rearranged itself into a pattern of nodes separated by only half the previous distance. This seems like a very clear demonstration of the 2 : 1 ratio in connection with the octave. And as for the reason that the musical fifth sounds so satisfyingly harmonious, Galileo imagines two strings vibrating, one of them making three vibrations for every two of the other. Every second beat of the slower string coincides with a beat of the faster string, but in between are beats that subdivide the beats of the faster string, and this rapid and playful alternation in behavior must somehow be delicious to the ear, Galileo suggests, "like a tender kiss and a bite."

Galileo's speculation about how the 3 : 2 consonance affects the ear is right in the spirit of the old theory that musical consonance depends on the integers in the ratio being *small* integers. The musical fourth, at 4 : 3, uses bigger integers, and so it is not quite as consonant as the musical fifth 3 : 2, which in its turn is not as consonant as the octave 2 : 1. The major third is traditionally the ratio 5 : 4, but in the strictest Pythagorean theory this is at best an "imperfect" consonance, because it involves the integer 5, which is getting to be on the large side. The major sixth, at 5 : 3, is also an imperfect consonance in this way of thinking.

We can visualize these intervals in terms of familiar musical instruments. On the piano it is convenient to think of the white keys starting at middle C and going up the scale to the C an octave above. This is the major scale, do-re-mi-fa-so-la-ti-do, CDEFGABC. The interval C-G, or do-so, is a fifth, and the interval C-F, do-fa, is a fourth. The interval C-E, do-mi, is a third. Alternatively, we can visualize the neck of a guitar, with its frets. A finger on the fret $1/5$ of the way along the string leaves only $4/5$ of the string free to vibrate, at the frequency $5/4$ higher than the open string, that is, at the musical third above the open string. If we put a finger $1/4$ of the way along the string, only $3/4$ of the string is free to vibrate, at the frequency $4/3$ higher than the open string, the musical fourth. If we put a finger $1/3$ of the way along the string, only $2/3$ of the string is free to vibrate, at the frequency $3/2$ higher than the open string, and we hear the fifth. And if we put a finger $1/2$ of the way along the string, we hear the octave. The neck of the guitar has frets at essentially these places (but not exactly, as described below) to make it easy to change the length of the string correctly before plucking it.

The intervals combine according to simple arithmetic. If we go up a musical fifth (factor $3/2$), and then go up from there by a musical fourth (factor $4/3$), then in all we have gone up by the factor

$$\frac{3}{2} \times \frac{4}{3} = \frac{12}{6} = \frac{2}{1},$$

that is, we have gone up by an octave. Combining intervals means *multiplying* the corresponding factors.

Now we can find the ratio that corresponds to the musical whole step F-G, also called a major tone. It is just the ratio that would take us from $4/3$ to $3/2$, namely

$$\frac{3}{2} \times \frac{3}{4} = \frac{9}{8}.$$

Since the integers 9 and 8 are rather large, we understand that the interval of the major tone 9 : 8 sounds rather dissonant, as F and G do when sounded together.

The scheme so far seems consistent and coherent, but here is where the trouble begins. The whole step F-G, at least on the piano or the guitar, looks like the same interval as other whole steps, like C-D, or D-E. Thus the major third, C-E should be two whole steps, or

$$\frac{9}{8} \times \frac{9}{8} = \frac{81}{64}.$$

Yet the major third was supposed to be $5/4 = {}^{80}/_{64}$, not ${}^{81}/_{64}$. The arithmetic would come out right if C-D were a major tone 9 : 8 while D-E were slightly smaller, a so-called minor tone 10 : 9, because

$$\frac{9}{8} \times \frac{10}{9} = \frac{10}{8} = \frac{5}{4}.$$

But is this what we actually do in practice? Do piano tuners really tune whole steps in two different sizes, 9 : 8 and 10 : 9? And if they do, how do they know which ones should be the minor tones, and therefore slightly flat?

There is really no simple answer to this dilemma. Not all the intervals of the scale that we normally use (which is also essentially the one that the Greeks of the Roman period used) can be the perfect Pythagorean intervals. There have to be compromises of the kind that we have hinted at above. Somewhere in the scale there have to be little adjustments, making intervals slightly sharper or flatter than they ideally should be. Insistence that some intervals be perfect makes problems for other ones.

Galilei's *Dialogue on Ancient and Modern Music* has just two characters, his patron Giovanni Bardi, to whom the work is also dedicated, and Piero Strozzi, a mutual friend and a member of the Florentine Camerata. On the question of tuning, Bardi asks Strozzi to tune a harpsichord by first making certain intervals perfect fifths, and to check the octaves that have thereby been determined.

> *Strozzi:* They agree excellently.
> *Bardi:* Proceed to tune the others corresponding to these.
> *Strozzi:* It's done.
> *Bardi:* Now play.
> *Strozzi:* This is music that really infuriates a gentle person who hears it, the kind Timotheus must have used to cause Alexander the Great to become enraged and take up arms. Although I hear the imperfect consonances as dissonant, I can't persuade myself how this comes about.[4]

Bardi had stipulated a very Pythagorean tuning, requiring perfect fifths, to the detriment of intervals like the third. And, in a sense, it seems to work. It is not, however, a tuning that would have been used in Galilei's day, or even, it appears, in polite company.

Galilei was an excellent lutenist and a competent composer. He had studied composition with Gioseffo Zarlino, later choirmaster at St. Mark's Cathedral in Venice, and he knew very well Zarlino's views on the question of scales and tuning. Zarlino had published his own book of music theory, advocating the syntonic scale from Ptolemy's *Harmonics,* a scale that has both major tones (9 : 8) C-D and minor tones (10 : 9) D-E, in the way that we introduced them above, and therefore a perfect third (5 : 4) C-E and also a perfect sixth (5 : 3) C-A. Some other intervals are necessarily compromised. For example, D-A is not a perfect fifth, because

$$\frac{5}{3} \times \frac{8}{9} = \frac{40}{27} = \frac{80}{54},$$

while a perfect fifth would have been $\frac{3}{2} = {}^{81}\!/_{54}$. Zarlino had nonetheless advocated this scale as the best possible compromise for modern music, and even claimed that this was the way that good musicians played and sang in actuality.

Galilei gradually parted company with his teacher on this question, and it is a theme of the *Dialogue* to point out that Ptolemy's syntonic scale, with its whole steps of two different sizes, cannot be the scale in actual use, and that it is in fact unworkable for some instruments. Galilei remains very respectful of Zarlino throughout, at least when he mentions him by name, and explicitly praises his contributions to music theory. Intellectual disputes in the Renaissance were never without fireworks, though, and finally in the dedication to Bardi, which was written last, Galilei boils over:

> If my work does not reach you [Bardi] with the refined prose that it should have . . ., certain unreliable Venetian printers are to blame, who held it up for many months—without any right— to please someone [Zarlino] who prevented my efforts from seeing the light, led either by envy or the wish to honor himself with my long labors [i.e., to plagiarize it]. They went so far as to deny me my original, which I had sent them already last October to publish. Before I could recover it, for the reason mentioned, I had two-thirds of it published [here in Florence] from a draft that I retained.
>
> On this I close, and with every reverence I kiss your hand.[5]

An old Renaissance joke says that the musicians spent half their time tuning and the other half playing out of tune. Galilei points out that this is unavoidable if Zarlino's scheme is adopted, probably enough by itself to account for Zarlino's attempt to suppress the book. A harpsichord can be tuned according to Ptolemy's syntonic scale, but fretted instruments like viols and lutes cannot. Their intervals are built into them in the fixed placement of their frets, which occur at fixed ratios of the total length of the string, the same for all strings. A major tone on one string requires that the same fret give a major tone on all the other strings, even if Ptolemy's syntonic would require on some other string a minor tone, not a major one. In practice what musicians do—and here Galilei would have known very well what he was talking about—is split the difference, using neither major nor minor tones, but making all the whole tones the same. For trying to harmonize with other instruments, the best compromise, far from ideal, is just to make the notes sound as nearly the same as possible, and this has nothing to do with small integer ratios. In fact, in the tuning of lutes and viols there are no perfect fifths. The fifths are slightly flatter than perfect, and the fourths are slightly sharper.

If the viols cannot tune to the harpsichord, perhaps the harpsichord could tune to the viols, and we could forget about perfect intervals. (In effect, that is what we do now.) Galilei considers this possibility but rejects it. He says that the harpsichord would then sound out of tune, because we are accustomed to hear perfect intervals from it. The sharp plucked sound of the string is much less forgiving than the softer note of the viol or the lute. Thus, for many reasons, different types of instruments cannot be in tune with each other. They use subtly different scales. And this is just one of the problems of modern music, he contends.

Monody

Vincenzo Galilei was an expert in polyphony, and knew how to score music for the lute in several parts, so that the notes harmonized together, just as soprano, alto, tenor, and bass harmonize together in vocal music. He even wrote books about this. But in the *Dialogue on Ancient and Modern Music* he blames polyphony for the sorry state of the music of his time. For one thing, the tuning problem arises unavoidably because notes must be harmonious when they are sounded together. This requirement is a severe constraint on tuning. The chords of the polyphonic harmonies must be present in the scale. Greek music, on the other hand, was not polyphonic. It had no need to be harmonious in the sense of modern music. The notes were never sounded together, only one at a time, in melodies. Greek music was monody, a single voice, accompanied by a single instrument, usually a lyre. This changes the problem of scales and tuning completely, and allows much more freedom.

Not only does polyphony put music into a straitjacket so far as tuning goes, but in a setting like, say, a four-part madrigal, polyphony completely obscures the words. The singers may even be singing different words at the same time, a veritable babble, and the different lines will almost certainly be doing different things, some going up, some down, all contradictory. How different from monody, in which the music and the words agree, and combine clearly and perfectly!

One might object that the Italian madrigal is a delightful musical form. Galilei did not deny this, and he never stopped writing polyphony himself, but he considered the music of his time *merely* beautiful, produced by composers who vied for the approval of an audience that wanted only entertainment. Their idea of integrating music and words went no further than such obvious devices as an ascending line for a happy subject, a descending line for a sad subject, dissonance to express torment, and so on. Galilei was sure that music could be more than this. He did try writing monody, composing a *Lamentation of Jeremiah* and a setting of a scene from Dante's *Inferno.* These compositions were performed in the Florentine Camerata, but they do not survive.

Galilei died in 1589, and the Florentine Camerata ceased to meet, but the idea of monody lived on. In 1600 the singer and composer Jacopo Peri wrote monodic music for an entire dramatic production that does survive, *Eurydice.* It was performed for an elite audience in the Pitti Palace in Florence as part of the celebration of the proxy marriage of Marie de Medici to Henry IV of Navarre, making Marie queen of France. The title *Eurydice* is perhaps an allusion to Marie, since it should really have been called *Orpheus,* one would think. In any case, the myth of the singer who could move even the stones and the trees by his music and who won back his wife from the underworld through the power of his art was the perfect vehicle for the celebration of a new, Greek-inspired musical style. Orpheus sings happily with his friends, the shepherds, in celebration of his wedding to Eurydice. The terrible news of Eurydice's death by snakebite and Orpheus's descent into the underworld are as dramatic as anyone could wish. As the story usually goes, the return from the underworld is permitted, but only under one terrible condition: Orpheus may not look back at Eurydice until they have ascended. And as everyone knows, that is impossible—he does look back. In Peri's version, though, the gods are unaccountably permissive. Eurydice returns unscathed, and she and Orpheus live happily ever after. A ludicrous defanging of the old myth, perhaps, but it was composed for a wedding.

At almost the same time, a similar production, *Orfeo,* by Claudio Monteverdi, partially inspired by Peri's *Eurydice,* was performed at the court in Mantua, where it was wildly successful. Clearly there was a market for more of these monodic musical dramas, and they kept coming, soon staged in their own theaters, beginning in Venice. They came to be called *opera.*

Galilei and his contemporaries believed that the ancient Greek tragedies were sung, not declaimed, and more recent evidence tends to confirm this idea. In the twentieth century fragments of music from two plays of Euripides, *Orestes* and *Iphigenia at Aulis,* were discovered on papyrus.[6] In any case, the hypothesis that sung plays would have remarkable power has been thoroughly vindicated—the history of opera proves that. Modern opera must use the modern scale, though, and that is essentially, up to subtleties of tuning, the scale described by Ptolemy, postdating the Roman conquest. No one ever claimed great power for Greek music of the Roman period, but the ancient tragedies were centuries earlier, pre-Hellenistic. What was that music? And what was the real music of Orpheus?

Ancient Scales

Galilei, like Zarlino, relied on the late-classical music theorists Ptolemy and Nichomachus (as transmitted by Boethius), but he gradually became aware of the importance of the much earlier theoretical writings of Aristoxenus, a pupil of Aristotle, fourth century B.C.E. These writings are fragmentary (only three short books survive), but they paint a very different picture. Other fragments from this period, and even earlier, confirm the impression, and the late theorists Ptolemy and Nichomachus also preserve a memory of the old scales. Most of the scales described by Aristoxenus, with names like "chromatic" and "enharmonic," are as different from the modern Western scale as Indian ragas are. (It should be noted that Indian music remains monodic to this day, in a continuous tradition going back to the time of Aristoxenus, or even earlier. Did Greek music sound like that?)

The scales of Aristoxenus are organized into two *tetrachords*, with the extremes of a tetrachord separated by a musical fourth, like E up to A, for the lower one and B up to E for the higher one, with a whole step A-B separating them. It was also possible for the higher tetrachord to start where the lower one left off, namely A, so that the extremes of the higher tetrachord would only be A up to D. We could get the usual diatonic scale if we filled in the interiors of the tetrachords with the usual white notes of the piano, creating among others the imperfect consonances of the third and the sixth. But Aristoxenus describes filling in each tetrachord with two pitches that may be in quite unexpected places, with intervals as small as a quarter tone. The actual positions of these interior pitches are moveable, within limits, giving a wide spectrum of possible scales. If, to take an extreme case, we think of the two interior pitches as contributing just one real pitch, with the other being more of a microtone ornament, and if the two tetrachords share the middle note, then there are really just five pitches in the scale instead of eight, a pentatonic scale, common in folk music the world over.

Galilei first mentions Aristoxenus in connection with the problem of tuning lutes. We have seen that two major tones (9 : 8) do not make a perfect third (5 : 4). The tones are just a little bit too large. Ptolemy introduces the minor tone (10 : 9), because a major tone (9 : 8) followed by a minor tone (10 : 9) is exactly a perfect third (5 : 4), but Galilei has argued that lutes can't do that. Their tones must be all the same. Now if two major tones are too big, it is easy to see, by a similar computation, that two minor tones are too small to make a perfect third. Ptolemy argues, in favor of his syntonic scale, that it is therefore impossible to make a perfect third with two equal tones. Neither choice works. But Aristoxenus's approach is different, and it is just what Galilei says musicians actually do: they do it by ear. If you start with a perfect third, you can find an interior pitch that splits the interval in half. You don't have to describe it by an arithmetic ratio. Intervals are not ratios, they are things that you hear. If you insist on giving a ratio, you can just measure where lute-makers put the frets, and you find out that they are located not quite at the traditional Pythagorean positions.

It had to be pointed out that music is not arithmetic. This startling fact, which is really the point of Galilei's *Dialogue*, is an indication of how seriously philosophy had intruded into music. For Ptolemy, and perhaps for Zarlino, the musical intervals *are* ratios. Ptolemy was ready to prove things about music by arithmetic, even things that made no sense in terms of actual musical practice.

Between the time of Aristoxenus and the time of Ptolemy, the role of mathematics in music theory had changed completely, in ways similar to what we have noticed before in arithmetic. For Ptolemy mathematics is a part of philosophy connected to the heavens, and to the heavenly harmonies (he even relates the notes of his scales to the planets). The musical notes of the syntonic scale have to correspond to small integer ratios, not because of the way they sound, but to fit a philosophical preconception. By an odd coincidence this choice of philosophical organizing principle probably *did* enable the (much later) development of polyphony, since the notes of such a scale furnish many harmonious consonances. This was not at all the reason for choosing the syntonic scale, though. The reason for choosing it had nothing to do with how notes sounded together, because they were not sounded together. The syntonic scale was prescribed by a philosophical idea, the same idea that links music and arithmetic together in the quadrivium.

By contrast, Aristoxenus uses mathematics to describe what musicians actually did and what scales they used. There is no suggestion that mathematics has anything to say about which scales are superior to others. Aristoxenus's use of mathematics is essentially scientific, not philosophical, and not judgmental. He uses it to describe what he hears:

> Some of our predecessors introduced extraneous reasoning, and rejecting the senses as inaccurate, fabricated rational principles, asserting that height and depth of pitch consist in certain numerical ratios and relative rates of vibration—a theory utterly extraneous to the subject and quite at variance with the phenomena . . . Our subject-matter then being all melody, whether vocal or instrumental, our method rests in the last resort on an appeal to the two faculties of hearing and intellect. By the for-

mer we judge the magnitudes of the intervals, by the latter we contemplate the functions of the notes.[7]

This, it should be pointed out, is also what Galilei does with mathematics. He does not argue mathematically for the superiority of one system or another, as Zarlino and other music theorists before him had done. Rather, he seeks to clarify the phenomena of music, as he knows them by experience, using mathematics as one of several tools. This is quietly to dethrone mathematics as a static element of philosophy, and to turn it back into something active.

It would be fair to call Ptolemy and Nichomachus Neoplatonists. Their preoccupation with small integer ratios in music as expressions, somehow, of what is good and right is something that Plato himself addresses in *The Republic*. Socrates and Glaucon are not satisfied to see their students investigating music in a merely scientific way. The vignette below shows how people like Aristoxenus must have appeared, carefully using their ears.

Socrates: We must constantly hold by our principle, not to let our pupils take up any study in an imperfect form, stopping short of that higher region to which all studies should attain, as we said just now in speaking of astronomy. As you will know, the students of harmony make the same sort of mistake as the astronomers: they waste their time in measuring audible concords and sounds one against another.

Glaucon: Yes, they are absurd enough, with their talk of "groups of quarter-tones" and all the rest of it. They lay their ears to the instrument as if they were trying to overhear the conversation from next door. One says he can still detect a note in between, giving a smallest possible interval, which ought to be taken as the unit of measurement, while another insists that there is now no difference between the two notes. Both prefer their ears to their intelligence.

Socrates: . . . they are intent upon the numerical properties embodied in these audible consonances: they do not rise to the level of formulating problems and inquiring which numbers are inherently consonant and which are not, and for which reasons.

Glaucon: That sounds like a superhuman undertaking.

Socrates: I would rather call it a "useful" study; but useful only when pursued as a means to the knowledge of beauty and goodness.

Glaucon: No doubt.[8]

Mathematics, Music, and Uncertainty

What the "absurd" pupils were doing has an interpretation in modern scientific terms, whether or not they would have thought of it this way. They were establishing the uncertainty in their measurements. The ear is a very good measuring instrument, but like all instruments it has its inherent limits. How different does a tone have to be before we perceive it as different? What is the uncertainty in the measurement? That is a fundamental thing to keep in mind in assigning numerical values to notes.

Since the ability of our ears to discriminate notes is a fundamental fact about how we perceive sound, let us estimate it. The absurd pupils are apparently talking about quarter tones as sounds that they can hear very well. Since a whole tone 9 : 8 is a factor $9/8 = 1.125$ in frequency, that is, a 12.5 percent increase in frequency, a quarter tone would be about a 3 percent increase. Similarly, the difference between a major tone 9 : 8 and a minor tone 10 : 9 is a tiny interval called by Ptolemy the "comma," $(9/8) \times (9/10) = {}^{81}/_{80} = 1.0125$, which is a 1.25 percent change in frequency. The discrepancy between harpsichords and lute tunings in Galilei's day was roughly half this size, and it was noticeable enough to make harpsichords and lutes not agree very well. Thus, it is clear that when we estimate pitches carefully, we are making a measurement with an uncertainty of about 1 percent, or perhaps a bit better than that in the case of a trained musician. There are many quantitative sciences today where a measurement good to 1 percent would be considered excellent, not usually attained. That is, the absurd pupils were doing quantitative science at a very respectable level of accuracy.

According to Plato, the people who do this sort of careful measuring are Pythagoreans. Thus the Pythagoreans were apparently working to understand music and the sense of hearing, not to discover the Good and the Beautiful. That is what makes them absurd in Plato's sense. Pythagoras, according to stories, aimed to understand music, and did understand it, without prejudice. When he heard the harmonies in the smith's shop, he rushed in to measure the hammers. He did not declare some of the hammers bad. He did not rail against some modes and in favor of others, as Plato does. Rather he *understood* the modes and used them "to restore the pristine harmony of the soul." The Taoromenian youth shouldn't have been listening to the Phrygian mode, but it wasn't because Phrygian was bad. It was just inappropriate at that moment. Pythagoras used his knowledge benevolently, but that is not the same thing as learning benevolence from music. Pythagoras's knowledge of music is a workmanlike knowledge, and any ethical use of this knowledge is motivated from elsewhere, not from within music. Clearly such knowledge can also be abused, as the Phrygian piper unwittingly did.

A similar workmanlike attitude must have characterized the building of the first pipe organ by the Hellenistic engineer Ctesibius. He would have found out all about the tolerances of the ear in tuning the organ. Cutting pipes according to theory, with integer ratio lengths, would not have worked. The lengths had to be adjusted. Pipes in the wrong ratio were still good, not bad, if they sounded right. Their ratios were irrelevant.

Pythagoras discovered the small integer ratios in music, but it is not at all clear that he assigned moral significance to them. Plato does that, in the voice of Socrates, but it still does not appear that this moral viewpoint claimed any particular following during the Hellenistic period. Aristoxenus, in fact, ridicules this idea. Then, in the Roman period, music became moral in the same way that arithmetic became moral. It was this peculiar moralizing role for mathematics that was to come undone again, at least partially, at the end of the Renaissance. In the work of Galilei, who paid close attention to the musical practice of his time and thus rediscovered the relevance of the earlier theory of Aristoxenus, we see how work in the arts pushed the meaning of mathematics back to what it had been in the Hellenistic period, more like the meaning that it has for us now.

When Socrates mocks the pupils, we have both meanings of mathematics before us, mathematics as scientific tool, in something like the modern sense, and mathematics as conduit to the Good. Predictably, Socrates regards mathematics as a conduit to the Good, but is Plato perhaps misrepresenting him here? There are intriguing little indications that the real Socrates might actually have been more scientific than Plato makes him appear. In *The Clouds,* an early comedy by Aristophanes, a younger Socrates is a comic character who is exactly like one of the absurd pupils in the *Republic.* He studies things up close, in the most minute and quantitative way, and is entirely intent on getting the right measured number, without any preconception of what it should be. He keeps a school called the Thinkery, and the wily old man Strepsiades, who is considering enrolling at the Thinkery, manages to find out from a student what sort of thing Socrates does.

> *Student:* Listen. Just a minute ago Socrates was questioning
> Chairephon about the number of flea feet a flea could
> broadjump. You see, a flea happened to bite Chairephon
> on the eyebrow and then vaulted across and landed on
> Socrates' head.
>
> *Strepsiades:* How did he measure it?
>
> *Student:* A stroke of absolute genius. First he melted some wax.
> Then he caught the flea, dipped its tiny feet in the melted
> wax, let it cool, and lo! little Persian booties. He slipped
> the booties off and measured the distance.[9]

If this scientific Socrates is hard to recognize, recall that just before drinking the hemlock, in Plato's *Phaedo*, Socrates tells his students about his early interest in the sciences and physical investigations, and his subsequent disillusionment with "gases, liquids, and other absurd things." Perhaps this scientific period in Socrates's life was actually longer and more important to the real Socrates than Plato lets us know, in creating, as he does, the later Socrates, exclusively intent upon the Beautiful and the Good. The Socrates of Plato's *Apology* even tries to explain away the scientific Socrates of *The Clouds* as an incomprehensible invention, but it is not completely convincing.

Vincenzo Galilei was not ashamed to emulate the absurd pupils in doing real experiments and making real measurements. His son Galileo was around twenty by this time and must have participated in the investigation. They repeated the experiments attributed to Pythagoras, taking identical strings, putting them under tension with various weights, and measuring the musical note produced by the string when plucked. They found that it was not true that doubling the weight made the note go up an octave, although this had been music dogma for nearly two thousand years. They had to quadruple the weight, not double it. All the Pythagorean ratios were found, except that the weight had to go as the square of the ratio, not the ratio itself. To raise the note by a musical fifth 3 : 2, for example, they had to increase the weight by the factor $(3/2)^2 = 9/4$. In all this time the physical reality of the musical ratios had apparently not been of too much concern. It was all a convenient philosophical theory, and the fact that it was wrong in certain important places was so unimportant that it had not even been noticed. There can be no clearer proof that mathematics in this period was not actually being used in the way that we use mathematics now. It served a completely different function.

Fifty years later Galileo included these experimental results in *Two New Sciences*. A bare account of the experiment makes it sound as if the numbers appeared easily and unambiguously once someone finally decided to measure them. If you try it yourself, though, you find, as you find in any experiment, that measurement uncertainty complicates the interpretation considerably. To see the result convincingly requires good experimental judgment.

Throughout his career Galileo had excellent taste in making such experimental judgments, a subtle prerequisite for establishing the link from mathematics to the world. His first encounter with the issue of measurement uncertainty must have been this experiment in music. Later, in his own *Dialogue* on the Copernican and Ptolemaic systems, he considers the comparison of theory and experiment in a homely metaphor:

> Just as the merchant who wants his calculations to deal with sugar, silk, and wool must discount the boxes, bales, and other packings, so the mathematical scientist, when he wants to recognize in the concrete the effects which he has proved in the abstract, must deduct the material hindrances, and if he is able to

do so, I assure you that things are in no less agreement than arithmetical computations. The errors, then, lie not in the abstractness or concreteness, not in geometry or physics, but in a calculator who does not know how to make a true accounting.

Galileo does not tell us exactly how to make a true accounting, to see how, if at all, the abstract and the concrete, the theory and the experiment, agree. Indeed, it is part of the education of every scientist to learn, through experience, what this means.

It is significant that a good musician can make accurate quantitative measurements without any special scientific measuring instrument, because the human ear is already such an instrument. Galilei, through careful measurement, had found startling weaknesses in Neoplatonic conceptions of music. By the end of the *Dialogue* it would be consistent to say, although Galilei does not come right out and say it, that small integer ratios, except for the octave, are irrelevant to music. Going further, early Greek music was powerful precisely because it was *not* constrained to be Neoplatonic and to stick to small integer ratios, even approximately.

Because of the limitations of the ear, it does not make sense to describe musical intervals like $^{10}/_9 = 1.111 \ldots$ to much more than three decimal places or so, but it is interesting to notice that the equal tones that Galilei says are in actual use are *irrational* ratios, like the numbers that Piero della Francesca was experimenting with in his *Libellus*. If a perfect third, $5 : 4$, is divided into two equal tones, then each of them is the square root of $^5/_4$, which is the irrational number $\sqrt{5}/2 = 1.118 \ldots$, a little smaller than $^9/_8 = 1.125$, and a little bigger than $^{10}/_9$. A few years after the publication of the *Dialogue*, the Dutch engineer Simon Stevin proposed the equal tempered scale of twelve equal semitones, each the twelfth root of 2, an irrational number that is about $1.0595 \ldots$, so that 12 semitones would exactly make an octave. In the equal tempered scale the tone would be two of these semitones, about I.1224 . . ., slightly less than the perfect 9 : 8 tone. One does not have to follow these calculations in detail to know that music theory was being turned on its head. Before Galilei, everyone knew that musical intervals corresponded to certain very special rational numbers. After him it began to look as though they shouldn't be rational at all.

8 Kepler and the Music of the Spheres

Johannes Kepler was interested in music too, but for a very different reason. The music of the spheres, the music of the universe as a whole, was somehow the key to understanding the cosmos: Kepler was sure of that. And with his unparalleled ability to reduce large quantities of observational data to harmonious order, he strove to apply music theory to astronomy. His discovery that Ptolemy had done the same thing in the *Harmonics* fired his eagerness. In Kepler's own voice from his 1619 *Harmony of the World*,

> My appetite was particularly intensified and my purpose stimulated by the reading of the *Harmonics* of Ptolemy . . . There I found unexpectedly, and to my great wonder, that almost the whole of his third book was given up to the same study of the celestial harmony, one thousand five hundred years before . . . This identity of conception, on the conformation of the world, in the minds of two men who had given themselves wholly to the study of nature, was the finger of God, to borrow the Hebrew phrase, since neither had guided the other to tread this path. Now, eighteen months after the first light . . . it is my pleasure to yield to the inspired frenzy . . . See, I cast the die, and I

write the book. Whether it is to be read by the people of the present or of the future makes no difference: let it await its reader for a hundred years, if God Himself has stood ready for six thousand years for one to study Him.[1]

This grandiosity must be forgiven, because Kepler really had done something amazing. The *Harmony of the World* contains the announcement of Kepler's Third Law, which, together with his earlier results, is a true description of the solar system, the end of a search that had gone on for two thousand years. And it really was music that had inspired him. In this work on Copernican astronomy one might naively expect Kepler to make some reference to Galileo, but he does not. He does, however, refer repeatedly to Galileo's father and the *Dialogue on Ancient and Modern Music*.

The First Exchange of Galileo and Kepler

Galileo's great contemporary Johannes Kepler was born in 1571 in Weil der Stadt, Württemberg, and died at the age of fifty-nine in Regensburg when Galileo was sixty-seven. They were aware of each other, and sometimes in correspondence, for most of their professional lives, and the nature of their relationship has puzzled historians ever since. In principle they should have been very close. In retrospect these two, without competing, were the great astronomers of their time. Both of them championed the cause of mathematics in the description of the world, both championed Copernicus, and in the final summation both are cited by Newton in his eventual synthesis of the laws of motion. In view of all these superficial similarities, their differences (and they are great) are most revealing. These differences seem to have puzzled even their contemporaries. In a letter of 1634 Galileo insists upon putting as much space as possible between himself and Kepler, years after the great man's death: "I have always esteemed Kepler for his free (perhaps too free) and subtle genius. But my way of philosophizing is completely different from his, and although it may be that in writing about the same material (and that would only be the motion of heavenly bodies) we have sometimes arrived at similar concepts, in that we assigned to some true effect the same true cause, you will not find this in more than one percent of my thought."[2]

The only unreservedly cordial letter Galileo ever wrote to Kepler was his first, in 1597, thanking Kepler for his book, but before he had read it.[3] He was writing in haste because Kepler's associate Paul Hamberger, who had brought the books (two copies of Kepler's *Mysterium Cosmographicum*), could not stay and was about to depart for Germany. Galileo comments only on Kepler's introduction, a synopsis of the Copernican theory that was not really representative of the book as a whole. Kepler had only included it there at the suggestion of his teachers at Tübingen. With reference to this first chapter Galileo avows that he too is a Copernican, the first indication of that kind that we have from him. Galileo apparently did not have time to digest the book's novel thesis, beginning in chapter 2, nor to comment on it in his letter.

Kepler was delighted with this letter, and replied six weeks later in a long, affectionate letter that Galileo must have found over familiar in the extreme.[4] Kepler was eager to have Galileo's criticism of the rest of his book, now that he had presumably read it. And as for Copernicanism, Kepler admonishes, "Be confident, Galileo, and go forward," in response to Galileo's avowed reluctance to publish Copernican ideas, being unwilling to endure the ridicule of the ignorant "whose number is infinite." Kepler recognized a typically Pythagorean attitude, but was eager for Galileo to join him in the public campaign for Copernicanism. "And if Italy is not so appropriate for your publications, and if something is holding you back, perhaps Germany can grant you that freedom." This sentiment must have irritated Galileo no end, even if, ironically, at the end of his life he would pursue this strategy with *Two New Sciences.* In truth, as of 1597 he had not published anything at all, and would not for many years to come.[5] "But enough of this," says Kepler. "You may write to me in private, at any rate, if you do not wish to publish, if you have found something pleasing in Copernicus." Worse and worse! This officiousness on the part of an obscure German Protestant, whom he had too unguardedly taken into his confidence, was quickly becoming tiresome.

In the same letter Kepler proposes an astronomical observational program that he and Galileo will undertake together, mentions that he is sending two *more* copies of his book, and closes with the anticipation of the very long letter that he will be getting in reply. What he got was—nothing.

When Galileo opened Kepler's book he found the most daring Neoplatonic document since *Timaeus* itself.[6] It proposed that the plan of the universe as a whole was based on the Platonic solids nested into each other. The structure of Creation could not be an accident, Kepler argued. The Creator must have used geometry to order the planets in their orbits. The various spheres on which the planets move had sizes that must have some geometrical meaning, and they originated in some deep and natural principle. In Kepler's mind this had to be the Platonic solids, because these fascinating structures could indeed determine the sizes of the spheres. For each solid one could imagine the inscribed sphere, like a balloon blown up inside until it just touched the faces, and the circumscribed sphere, like a balloon outside that shrinks until it just touches the vertices. The radii of these two spheres were just what Piero had been computing in his *Libellus* more than one hundred years before, and the corresponding ratios are mysterious natural proportions pre-existing in mathematics that the Creator would have had available to Him to order the world.

Once the idea comes to mind, its execution is almost obvious. The large gap between Mars and Jupiter is a good place to start. The sphere of Jupiter is more than three times bigger than the sphere of Mars. Why should there be such an enormous void? Well, if we inspect the Platonic solids, we find that one of them, the tetrahedron, has a circumscribed sphere almost three times larger than its inscribed sphere. That's it: Jupiter and Mars are separated by the tetrahedron. Not only that, but with five Platonic solids in all to separate the planets we determine six spheres, and there are exactly six planets. Amazingly, the ratios of the planetary orbits to each other are tantalizingly close to the ratios from geometry, close enough that Kepler had no difficulty deciding which Platonic solid separated each adjacent planetary pair, as the tetrahedron so obviously separated Mars and Jupiter.

Kepler could not have imagined how crackbrained this all sounded to Galileo.[7] He would have been astonished to learn that Galileo, despite his declaration in favor of Copernicanism, was not particularly interested in Copernicanism, and was not even particularly interested in astronomy. As for joining Kepler in an observational program, Galileo did not do observations.[8] Galileo had to teach astronomy only because he was professor of mathematics, but he had come to mathematics from a completely different direction, as we have seen. He must have learned astronomy on the job, letting it interfere as little as possible with his real interests.

Kepler and Tycho

Kepler, on the other hand, had come up out of poverty, pursuing a conventional education in mathematics and astronomy as a philosophical foundation for becoming an ordained Lutheran minister, an ambition that he was never to realize. In the end he came to believe that God had found in astronomy a better way for him to serve. Unlike Galileo, Kepler revered his teachers, especially Professor Michael Maestlin at the University of Tübingen, from whom he had learned his

Copernicus. Maestlin supported him in the publication of *Mysterium Cosmographicum*, and even assisted with some of the computations,[9] an astonishing idea to anyone who knows Kepler's own stamina for computing. Maestlin and Kepler must have been very much alike, except that no one could match Kepler's daring freethinking in the Neoplatonic manner.

It was undoubtedly Kepler's enthusiasm for numerological and geometrical speculation that scotched his plans for the ministry, because when the faculty at Tübingen got a request from their school in Graz, Austria, for a mathematics teacher, they sent Kepler. When he wrote to Galileo, he was outwardly a twenty-six-year-old high school teacher of mathematics, but inwardly a Neoplatonic visionary with a private conception of the world that had first struck him one day in the classroom and that never left him. Twenty-five years later, when he had made all the great discoveries for which we remember him, he issued a second edition of *Mysterium Cosmographicum* in which he changed nothing in the original text except to add notes to show how all his subsequent works had only filled in the places prepared for them in this, his seminal book.[10]

Besides Galileo, Kepler had eagerly sought comments on his book from Tycho Brahe, the renowned Danish astronomer on his island of Hven, a virtual fiefdom held from the king of Denmark, dominated by Tycho's fantastic residence and observatory, Uraniborg. Like Galileo, Tycho made a friendly reply, pointing out that the data Kepler had used for the planetary orbits, taken from Copernicus, really depended on the same observations that Ptolemy had used, and that he, Tycho, had modern data that were much more accurate. Kepler gradually came to appreciate, over the next several years, that the Prutenic Tables, a reduction of Copernicus's method to workaday form, could miss the true position of Mars by over four degrees, a huge discrepancy (eight times the width of the full moon), so that nothing in traditional astronomy, or even Copernican astronomy, was reliably accurate. Already in 1597 he must have been intrigued by what Tycho was offering. To discern the real truth of the structure that he had so tantalizingly glimpsed, then, would require accuracy at the level of Tycho's data, but the chance that he would get access to those data must have seemed impossibly remote.

In that same year of 1597, though, a disaster occurred that would bring Tycho and Kepler together. Through a combination of carelessness and arrogance Tycho lost Hven and began a trek south through Europe with a baggage train of everything that he could move in search of a new patron.[11] Two years later, against all odds, he succeeded in getting the support of the Holy Roman Emperor Rudolph II in Prague, and began setting up his new headquarters. At the same time, Kepler had lost his job when the Catholic authorities closed the Protestant school. On the basis of a vague invitation, Kepler risked his professional life in going to Prague, where he succeeded in becoming Tycho's assistant. Within two years Tycho was dead, and Kepler, on the basis of his association with the renowned Tycho, was Imperial Mathematician to Rudolph II.

At this point Kepler should have used Tycho's data to figure out in complete detail how the Platonic solids determined the structure of the world, his original motivation for seeking out Tycho. The scheme was not perfect yet, and Kepler exercised his fertile imagination to find what was still missing. In particular he had determined that music had to play a decisive role, something that he had not yet included. And even prior to including music he would have to comprehend the data perfectly in order to see the pattern. In what must have been a maddening development, though, Tycho's heirs, especially his son-in-law Franz Tengnagel, took possession of the data, and for a time would not give Kepler access.[12] In the interim Kepler set himself to investigate a problem that was crucial to interpreting the data in any case, the problem of optics.

At the level of Tycho's accuracy certain optical effects, like refraction of light rays in the atmosphere, were much too large to ignore. Optical effects in eclipses and in the face of the Moon also fascinated Kepler. These things had been discussed in antiquity, not just in treatises on optics, reworked by the Arabs and their European translators, but also in Plutarch's *On the Face in the Moon,* a book that Kepler carefully translated. His *Optical Part of Astronomy,* published in 1604, is directed toward these topics, and seems to be unaware of artists' perspective theory.[13] It is possible, though, that he would not have been interested in such terrestrial notions. Kepler was mad for astronomy, and nothing else reached a comparable level of importance for him. Like the painters, but with a different motivation, he thought about how we actually see, and—let this be just one indication of his genius—he was the first to understand how images form on the retina. He described accurately how corrective lenses help the nearsighted and the farsighted, improving on what had been said a few years earlier by the Neapolitan Giovanni Batista della Porta in a book called *Natural Magic.* That remarkable book described effects with lenses, even including telescopes, but not very clearly or with any deep understanding.[14]

In describing corrective lenses, Kepler showed in the margin of the page two figures, one a concave lens and one a convex lens, with representative rays going through them.[15] The two figures happen to be one above the other along the same optic axis, so that if you take it to be one figure and not two, it depicts a Galilean telescope, something Kepler was to point out later. In truth though, as Kepler himself admitted, he was skeptical of telescopes until Galileo's discoveries. In his 1604 *Optics* he does not discuss lenses for astronomical use at all, despite discussing lenses and astronomy in the same book. In January 1610 when Europe was experiencing a sudden enthusiasm for telescopes, Emperor Rudolph suggested to Kepler that they have a look at the Moon through a small telescope. Rudolph was convinced that he could see the reflection of Italy and two nearby islands in the Moon, a notion reminiscent of Plutarch's dialogue. They never tried it, perhaps because Kepler discouraged the idea.[16]

When Kepler finally got access to Tycho's data again, he set himself to understand in particular the motion of Mars, since that planet showed the largest discrepancies with existing theory. Kepler's universe was an animate universe, in which the planets, including the Earth, were living beings. For the problem of planetary orbits this raised the question of how the planet would "know" how to steer through the vast featurelessness of the heavens. Both Ptolemy and Copernicus, in order to describe the epicyclic motions of the planets, had prescribed circles about geometrical positions where nothing was actually located, but this made no sense to Kepler. If there was nothing there, the planet couldn't see it, and couldn't use it as a landmark. What the planet could see, and steer by, was the Sun, and in a tour de force of dogged computation and inventiveness that has rarely if ever been equaled, Kepler found a far superior description of planetary motion in which the actual Sun, and not some geometrically constructed point in empty space, organized the motion most beautifully. He tells the whole story of this arduous investigation in his *Astronomia Nova* of 1609, truly a new astronomy, and one that we can recognize, despite its rather peculiar motivations, as our own.[17] And with this, as he tells us, "like a general who had won glory enough through a most strenuous military campaign, I took some respite from my studies."[18]

The Second Exchange of Galileo and Kepler

While Kepler published prolifically, Galileo still had not published anything worth mentioning, and he had snubbed Kepler's first attempt to set up a correspondence. Kepler seems nonetheless to have had a very high opinion of Galileo. This can only have been on the basis of indirect contacts, of which one was certainly Martin Hasdale, an Englishman, a friend of Sagredo and Galileo, an intimate in the Venetian salon society where Galileo shone so brilliantly. Hasdale had become Librarian to Rudolph II, and was therefore a colleague of Kepler's. Through Hasdale and others some indirect contact must have been maintained between Galileo and Kepler.[19] Thus Kepler did not need to be reminded of who Galileo was when his friend Councillor Wackher von Wackenfels pulled his carriage up to Kepler's house one day in April 1610, shouting that Galileo had discovered four new planets with a telescope. This sensational news, which most of the astronomers of Europe greeted with livid envy and disbelief, sent these two into paroxysms of joy, as Kepler tells it, laughing too hard to talk.

Kepler's first concern was for his model of the universe, which could only accommodate six planets, neither more nor less, so it was with great relief that he found in Galileo's book, *Sidereus Nuncius* or *Starry Messenger*, delivered a few days later, that these so-called planets were actually moons circling Jupiter. In this capacity they were most welcome. Perhaps they would explain some otherwise unaccounted-for space in the arrangement,[20] and they did not increase the number of actual planets. Kepler immediately conjectured other moons about other planets in a regular pattern that would require two moons around Mars, and around Saturn six or eight moons that he yearned to discover.[21]

The arrival of Galileo's book was accompanied by requests for Kepler's opinion about it from the emperor, from Galileo, and soon enough from other astronomers like Giovanni Antonio Magini at Bologna, who considered himself a rival of Galileo. Even now, though, Galileo had not exactly written to Kepler. Rather he wrote to the Tuscan ambassador, who relayed the request to Kepler, with the further request that Kepler reply in just seven days, in time for the next post. The reason for doing it this way was Galileo's skillful involvement of the Medicean bureaucracy in legitimizing his discovery as he parlayed the discovery of the Medicean Planets into a high position at the court of Cosimo II, leaving his university career behind.[22] Galileo desperately needed Kepler's affirmation, and Kepler, with characteristic generosity, obliged him. Not only did he reply in a long laudatory letter, but he quickly published it for everyone to see as the *Conversation with the Starry Messenger*.[23] As Galileo put it in a letter that he finally did write four months later, "I give you thanks that you were the first one, and almost the only one . . . that you had the ingenuity and sublimity of mind to put complete faith in my assertions."[24]

Kepler's rhetorical strategy in the *Conversation* had been to point out that although Galileo's discoveries were certainly astonishing, yet there was much that one could accept without difficulty. The mountains on the Moon, for example: as abhorrent as they might be to conventional Aristotelian philosophy, this was actually quite a familiar idea from Plutarch and also from Kepler's 1604 *Optics* (where he had drawn heavily on Plutarch). As for the construction of the telescope, one could look at della Porta's book, or Kepler's 1604 *Optics* again. And if so much was familiar and plausible, then why not accept the moons of Jupiter too? Galileo would certainly not want to ruin his reputation by dedicating imaginary planets to the Medici.

Kepler's own exuberant imagination was not to be repressed on the topic of what all this might mean, with his own speculation on the origins of the lunar craters. Like Plutarch, he enthusiastically imagined the inhabitants of the Moon:

> It surely stands to reason that the inhabitants express the character of their dwelling place, which has much bigger mountains and valleys than our earth has. Consequently, being endowed with very massive bodies, they also construct gigantic projects. Their day is as long as 15 of our days, and they feel insufferable heat. Perhaps they lack stone for erecting shelters against the sun. On the other hand, maybe they have a soil as sticky as clay.

Their usual building plan, accordingly, is as follows. Digging up huge fields, they carry out the earth and heap it in a circle, perhaps for the purpose of drawing out the moisture down below. In this way they may hide deep in the shade behind their excavated mounds and, in keeping with the sun's motion, shift about inside clinging to the shadow. They have, as it were, a sort of underground city. They make their homes in numerous caves hewn out of that circular embankment. They place their fields and pastures in the middle, to avoid being forced to go too far away from their farms in their flight from the sun.[25]

As for the moons of Jupiter, Kepler took that to mean that Jupiter too was inhabited, since someone must have been intended to enjoy the sight of those moons and feel their astrological influences.

In his eagerness to support the new discoveries and to seem a part of them by citing his own work, Kepler inadvertently gave arguments to Galileo's opponents, namely that there was nothing new in what Galileo had done, and rather that Kepler or della Porta should get most of the credit. And as for that one thing that was certainly new if it was not mistaken, the moons of Jupiter, they loudly denounced it as a fraud. For Kepler this was an agonizing experience, and he wrote long letters to Galileo, focusing in particular on one Martinus Horky, a student of Magini at Bologna, who had published an attack on the *Starry Messenger* sufficient to discredit Galileo with the grand duke, if it had been successful. Kepler describes how he had rebuked Horky in letters, reasoned with him in person when Horky came to Prague, and hoped to convert him to a more sensible position.[26] Horky meanwhile thought he might be converting Kepler.[27] These letters show Kepler truly distraught at the turn things had taken, his own words twisted against Galileo.

Kepler also described his own hapless attempts to make telescopic observations, begging Galileo to send one of his good instruments, or at least good convex lenses, which Kepler was unable to fashion himself. He complained of copyright infringements at Florence where his *Conversation* had been printed in a pirate edition, and he offered to take a good convex lens in reparation. To all of this Galileo replied only once, in August 1610,[28] suggesting that Kepler just ignore Horky. In that letter he also assessed what Kepler would and would not be able to see with the telescopes available to him, but did not promise to help him. By this time his appointment to the Medici court had come through, and Galileo's future was assured.

Kepler had held out long enough to accomplish what Galileo needed from him, but in fact the German astronomer was getting very nervous, and that too came through in his letters. He had approved Galileo's discoveries when he had no evidence for their correctness beyond what anyone else had, namely Galileo's book. If Galileo were wrong, Kepler too would suffer. And there were worrisome reports reaching him. On the evenings of April 24 and 25, Galileo with his telescope had stopped in Bologna on his way from Padua to Pisa, where he planned to show his new planets to their dedicatee, Cosimo II. On this stopover he meant to show them to Magini, the mathematician at Bologna, but the attempt was a disaster. According to Magini, neither he nor anyone else there, except for Galileo himself, could see the moons of Jupiter.[29] Normal stars appeared double in the telescope, convincing them all that you could not trust the instrument. Horky, citing a vague speculation of Kepler in the *Conversation*, later tried to explain what he had seen as just multiple reflections in the lenses.[30] What Kepler heard was that Galileo had left in great confusion. When Kepler begged Galileo to tell him who else had ever seen the moons of Jupiter,[31] the only person Kepler himself could name was Horky, and that was hardly reassuring. Galileo wrote that Cosimo II had seen them,[32] but that was hardly reassuring either. Some of Kepler's nervousness filtered through to the Magini camp, and it only encouraged them.[33]

Galileo's man on the ground in Prague, his friend Martin Hasdale, monitored what effect Magini was having on other people of influence, and he kept an eye on Kepler.[34] Magini was making a big show of being shocked! shocked! at the behavior of his student Horky,[35] but from Prague Hasdale reported his impression that Magini was actively working to undermine Galileo's credibility. In Hasdale's view, it was Rudolph II, the emperor himself, who had tamped down criticism of Galileo from that quarter.[36] The emperor was an astronomy buff—witness his patronage of both Tycho and Kepler—and Galileo had kept him very much in the loop, sending him updated observations and eventually a telescope, to which Kepler only got access secondarily.[37] Kepler meanwhile did his best to get his own telescope from Galileo, working now through the Tuscan ambassador Giuliano de Medici, to whom he had dedicated his *Conversation with the Starry Messenger*. Giuliano was very impressed with Kepler (whom he calls "Gleppero").[38] Kepler seems not to have realized that his best conduit to Galileo would have been Hasdale, who had befriended him, or at least buttered him up, although Hasdale had reported to Galileo, "I am not eager to ask him for one of his little books, since they have no merit, in your view."[39]

Everyone who looks at this story wonders why Galileo gave Kepler so little respect, not answering his letters or his pleas, and paying no attention to his momentous discoveries. When Hasdale wrote, casually acknowledging Galileo's low estimation of Kepler's books, Kepler had just published his *Astronomia Nova*, containing what are now called Kepler's First and Second Laws, the true description of planetary orbits. How could Galileo have been so blind that he didn't recognize Kepler's genius, either at this time or at any later time?

If one looks at what Galileo actually knew about Kepler, it is clear that Kepler had made a bad impression on Galileo right from the start, and it didn't get any better the second time around. Kepler contributed to this impression not just by his prolonged disorientation under the attacks of Horky and by his ineptitude as an astronomical observer, but by outright confessions of his failings. In a letter of December 1610 that he probably did not send, but which captures explicitly what is implicit in his other letters, he begins, "I, most brilliant Galileo, am not Italian, nor do I come from the most polite stratum of the German nation, nor was I brought up in an elegant home to use polite speech and gesture, such that I could match you in urbanity, distinguished astronomer" and confesses later, of his inability to make telescopes, "I am not good with my hands, given solely to speculations."[40] Perhaps even Kepler realized that such frank humility was not for Galileo, bred to a courtly culture that prized *sprezzatura*, making the difficult look effortless. But Kepler's feelings of inferiority were plain even when he didn't state them. In the next letter that he actually did send in January, 1611,[41] Kepler, who was continually hoping to participate in Galileo's discoveries, spent most of a page on fruitless attempts to guess Galileo's most recent one, which had been published as an anagram to be revealed later.[42] Although Kepler meant it playfully, it probably struck Galileo as pathetic. And Galileo, acutely sensitive to style, must have found Kepler's writing style overwrought and grotesque. There was virtually nothing in any of this to suggest to Galileo that Kepler had just solved the problem of the millennium.

Oddly, Kepler made almost no attempt to call Galileo's attention to *Astronomia Nova*. For a few tumultuous months he had Galileo's attention, but he mentioned the work only once, at the beginning of his *Conversation*, where he recalled how he had been dreaming that with the publication of *Astronomia Nova* he and Galileo might resume their correspondence, interrupted twelve years earlier. But no: Galileo has not been reading other people's books, he has been writing his own. A few pages further on in the *Conversation*, where Kepler imagined looking for his conjectured two moons of Mars,[43] he pointed out that Mars would be in optimal viewing position in October 1610, when the tables are in error by about 3 degrees. This would have been the ideal moment to emphasize that the tables were now obsolete—Kepler could predict the position of Mars at the level of accuracy of Tycho Brahe's observations. That, for practical purposes, was the point of *Astronomia Nova*. Unaccountably, he did not do so, except in this very oblique way. Perhaps Kepler himself did not fully appreciate his accomplishments. It is only in retrospect that we call them Kepler's Laws with capital letters.

Galileo almost certainly knew of Kepler's First Law, that the planets move in elliptical orbits with the Sun at a focus, but it is equally certain that he didn't believe it. It could have been just one more harebrained Keplerian fantasy as far as he was concerned. *Astronomia Nova* itself begins with an introductory section that is quite readable and entertaining, advising any non-Copernican who is "too stupid to understand astronomical science" to "betake himself home to scratch in his own dirt patch." But with these preliminaries over, it becomes a very forbidding book, concerned with subtle effects at a level of detail that Galileo was not even equipped to follow, never having made a serious study of astronomy.

Earthly Mathematics

Both Galileo and Kepler became court mathematicians, but their ambitions could not have been more different. Although higher mathematics in 1610 was still entirely identified with astronomy, Galileo meant to extend mathematics to earthly things, a philosophical revolution for which he needed the title Court Philosopher as much as Court Mathematician. For Kepler earthly mathematical problems simply could not compete with astronomical ones for beauty and importance. He says this explicitly in a note to the second edition of *Mysterium Cosmographicum*, summarizing his own life's work to that time, both astronomical and nonastronomical. "Heaven," he says, "the chief of the works of God, is much more notably embellished than the rest, which are paltry and mean." Thus, he says, his nonastronomical works "did not achieve anything which gave equal satisfaction" to his astronomical ones. Kepler, a prodigious mathematical talent, who actually did treat a few earthly topics with brilliance, still felt that such problems were "paltry and mean." His example tells us that Galileo's vision of an earthly mathematics on an equal footing with heavenly mathematics was a radical idea, and not at all obvious, even to someone perfectly equipped to understand it and even to participate in it.[44]

Kepler's most notable nonastronomical works are clustered, perhaps not coincidentally, in the few years following his second exchange with Galileo. The first of these, his *Dioptrice* (1611), is explicitly indebted to Galileo, quoting Galileo's letters (to other people) on his latest discoveries. Kepler returns in *Dioptrice* to a problem he had omitted from his 1604 *Optics,* the use of lenses in astronomy, and especially the theory of the telescope, since the telescope had turned out to be useful in astronomy after all. His treatment of this subject is a practical, semiquantitative account of what you see when you look through two lenses, undeniably the result of systematic experimental work, not something one usually thinks of in connection with Kepler.

Also in 1611, as a New Year's gift for his friend Councillor Wackher von Wackenfels, the same who had brought him the news of the Medicean Planets the year before, Kepler wrote an essay "On Six-Cornered Snow."[45] This little booklet, idiosyncratically playful in the Keplerian style, claims to offer its recipient a gift of Nothing, or as close to Nothing as he can find, a disquisition on the ephemeral snowflake, "the very thing for a mathematician to give, since it comes down from heaven and looks like a star." Kepler has a wealth of ideas to explain why snowflakes are hexagonal. Among his speculations are the hexagonal packings of spheres into planar and solid arrays, what we would now call crystallography, motivated by the packings of pomegranate seeds in their fruit and the cells of beehives. In what would be no surprise to his friend Wackher, Kepler considers the Platonic solids: one of them, the octahedron, has six vertices, so he devotes considerable space to why snowflakes should be octahedral. When they land, he posits, they collapse into a planar shape, still having all their six points. Alas, closer inspection shows that the flakes landing on his coat are already planar when they land, and not octahedral. The book is an entertaining tour de force of Neoplatonic whimsy, but it is a measure of Kepler's brilliance that modern readers cannot help seeing potential scientific insights in it. Kepler, however, insists that it is a joke.

Kepler's last nonastronomical book is the *Nova Stereometria Doliorum Vinariorum* (1615), new thoughts on the volume measure of wine casks. From the bibulous title this also sounds like a joke, and in a sense it is, since it was originally intended as a New Year's gift for a patron in 1613. Kepler became genuinely interested in the problem, though, and the final version, reflecting considerable additional work and thought, went far beyond his original conception.[46] He had it printed at his own expense, despite warnings that there was no market for such a thing, and even had a very arithmetical version printed in German the next year. One might erroneously assume that Kepler had conceived a sudden passion for practical commercial geometry. In fact, though, his intense involvement with this problem stemmed entirely from novel mathematics.

The loquacious Kepler tells us enough that we can reconstruct what happened. In buying wine for his cellar on the Danube landings in the fall of 1613, he was struck by the method used to determine the volume of the casks, and hence the price. The responsible official simply poked a graduated rod into the centrally located bunghole as far as he could, and read off the volume. Kepler reasoned that this method could not work, because one length measurement is not enough to determine a volume unless the casks had all been simply scale models of each other (geometrically similar), and they were not. Kepler might have felt that he, as Imperial Mathematician, had a responsibility to set things straight, and in this sense, he did perhaps conceive a fleeting interest in commercial mathematics. He had another reason for spending almost two years on this problem, however: he discovered that the usual method for measuring the casks *does* work, but only for a most subtle mathematical reason, and only for Austrian casks. Among all the wine casks of the world, therefore, Austrian casks are, in a certain practical sense, the best. The whole *Nova Stereometria,* including a first section on volumes of revolution in general, is motivated by this single point, and the mind-numbing computations of the German version, which superficially look like practical mathematics, are only recapitulations of the investigation that led Kepler to this insight and confirmed it.[47]

Kepler's Third Law

The German version of *Nova Stereometria* fulfilled its original function as a New Year's gift on January 1, 1616, and Kepler returned to astronomy, especially to the construction of new astronomical tables making use of his planetary theory. This work, the Rudolphine Tables, had been commissioned by the Emperor Rudolph, long since deposed by his brother Matthias and deceased, but Kepler worked faithfully nonetheless to honor his promise to his old friend and patron, finally publishing the tables in 1627. His inventive mind could not be content, though, with a perpetual diet of rote computations. He set himself, finally, to add the music to his planetary theory, the last step in his cosmography. It had become clear that the Platonic solids by themselves just set up the cosmic arena within which the celestial music was performed, and there were still gaps of understanding in the planetary arrangement. The question now was about the music. He studied Ptolemy, Galilei, and Zarlino, and with his characteristic exuberance invented a new foundation for musical consonances.

The old pairings of the quadrivium, geometry with astronomy and arithmetic with music, had to be unified in Kepler's scheme in order to bring music into astronomy. He did it by geometrizing music. The musical consonances, like 2 : 1, 3 : 2, including even the more imperfect consonances like the minor third, 6 : 5, never make use of the number 7. And when you look at the regular polygons that can be constructed by the methods of Euclid, like the square, the pentagon, the hexagon, what do you find? The seven-sided figure, the heptagon, cannot be constructed. Coincidence? Kepler didn't think so. Kepler had discovered that our experience of musical consonance is really an experience of geometry, and that this was a superior way of understanding it. When Galilei wondered why the fifth above the octave sounds more pleasant than the simple fifth, a question no doubt about the real experience of music in performance, Kepler found his answer "not worth examining" because his own geometric explanation was so much clearer.[48]

Kepler tried two ways to associate musical pitches with planetary motions. He considered that their pitches might be proportional to their speeds, or that they might be proportional to their *angular* speeds as viewed from the Sun. In any case, he knew from his own synthesis of the data exactly how they sped up and slowed down in their elliptical orbits due to their orbital eccentricities.[49] The resulting music is very peculiar, however. Two planets, Venus and Earth, with nearly circular orbits, change their pitches by roughly a half tone or less, as if they were just slightly tuning. Jupiter and Saturn range over intervals that could be almost as large as a major third. Mars could vary by almost a major fifth, and Mercury by more than an octave. As the planets speed up and slow down, they come into various harmonies with each other, always changing, and never resembling any known human music. In struggling to understand this situation, Kepler had to compare, to high accuracy, the properties of one planet's orbit with that of another, and that is how he discovered his Third Law, which, unlike the First and Second Laws, relates the orbits to each other: polyphony! Kepler's Third Law says that all the orbits are, in a sense, perfectly in tune, all agreeing with each other in a simple but unexpected way.

It must be said that this law does not suggest harmony in the way that Kepler or anyone else had anticipated. In various formulations it requires a square root or a cube root, and these roots were traditionally associated with irrational proportions, and hence with dissonance, not consonance. Kepler does not overtly express any discomfort over this, but the only example he gives to illustrate the law is very peculiar. The importance of the Third Law is that it is exact, as good as the data that Kepler used, but his example says that Saturn's orbital period is about thirty years, which is about twenty-seven years, and the cube root of 27 is 3. Squaring 3 we get 9, and hence the mean size of Saturn's orbit should be about nine times that of Earth, and this is roughly right. It looks as if he took the planetary pair that came the closest to yielding square roots and cube roots that were integers, since it was integer ratios that he wanted, but these approximations obscure what is most important about the law, its precise accuracy. Without doing his own computations, a reader of Kepler's book would have no idea how good it is.

In Newton's hands it would be revealed that the Third Law, never thought of in musical terms by anyone but Kepler, was the consequence of and the clearest indication for the inverse square law of universal gravitation. Thus there really is a music of the spheres, a music for the intellect, just as Neoplatonists had always maintained. And of all those who had ever strained to hear it, no one had ever heard that music so sweetly or so gratefully as Kepler, even if it wasn't quite the harmony that he expected.

Galileo's reaction, or rather nonreaction, to this discovery is telling. By 1619 he had been following the orbits of the moons of Jupiter for most of a decade in order to determine their orbital periods to exquisite accuracy. As part of this program he also had to know their orbital radii, in order to distinguish the four moons one from another at each observation. It is precisely these quantities, orbital radius and period, that Kepler's Third Law relates. As early as 1612, when he published preliminary values in the second edition of his treatise *On Floating Bodies,* Galileo already had all the data he would need to confirm the astounding fact that the moons of Jupiter also obey Kepler's Third Law.[50] Nothing reveals his dismissal of Kepler's Neoplatonism more clearly than his failure to check or acknowledge this possibility. It would have taken just a few minutes' computation, but he never bothered to do it.[51]

Architecture

The dome of Florence's cathedral Santa Maria del Fiore dominates the city today as it has done since it was completed in 1445. This architectural wonder not only eclipsed in size the Pantheon, which had been the largest dome in the world since Roman times, but it did so from a baseline elevation that was already 170 feet in the air. The man whose name has become almost synonymous with the great dome is Filippo Brunelleschi, the greatest architectural genius of the fifteenth century.

Brunelleschi left no written document on his method for designing the dome. Not even drawings survive. In painting he was one of the first to demonstrate accurate perspective constructions, but those paintings don't survive either. We are fortunate to have a small biography of Brunelleschi by Antonio Manetti written in the 1480s, some forty years after his death.[1] Manetti himself is called a "citizen and architect" in one city document. A member of one of its leading families, he served Florence in many government positions. Manetti had actually met Brunelleschi when Brunelleschi was close to seventy years old and Manetti was around twenty. In his biography Manetti claims to have seen and held in his hands the painting of the Baptistery with which Brunelleschi demonstrated his mastery of the method of perspective.

Manetti is also our source for the hilarious hoax perpetrated by Brunelleschi on one of his neighbors, a certain Manetto, in which he persuaded Manetto that he, Manetto, was not himself, but rather somebody else.

In recounting Brunelleschi's life, Manetti digresses to tell the story of architecture itself, how men first built crude huts from boughs and grass to escape the weather, how they progressed to crude stonework, to bricks and mortar, to dressed stone, and so forth. Nowhere does he mention the importance of mathematics in this process, and in fact history suggests that even monumental architecture requires almost no mathematics. Simple arithmetic and *geometria practica* seem to be enough. The ancient Egyptians had a reputation for being very mathematical, and the Greeks even said that they had learned mathematics from the Egyptians, but the evidence of papyri is that the Egyptians could not do much more than compute the number of bricks needed for a pyramid, or the number of loaves needed to feed the workers on a big project. By comparison, Babylonian arithmetic was far superior to Egyptian arithmetic, especially in its handling of fractions, but in the end this didn't seem to matter. Babylonian architecture was in no way superior to Egyptian architecture, and although Babylon and Egypt must have confronted each other in various ways over the centuries, superior mathematics never seems to have given the Babylonians a decisive advantage in any sense. A little arithmetic was enough, apparently, and more wasn't better.

Even the cathedral builders of the late Middle Ages still used only the simplest arithmetic and geometry. What they chiefly relied on was the wealth of accumulated experience passed down from master to apprentice, and this was still true in Brunelleschi's day. Brunelleschi solved many problems in building the dome, but he didn't use sophisticated mathematics to do it. The geometry of the interlocking pieces of stone, wood, and iron that distributed the enormous stresses of the dome was far too complex for the mathematics of the time to describe in detail—the design had to come from intuition and experience. When asked to put his plan into writing, in 1420, he was willing to put numerical dimensions on the components of the dome only up to a height of at most thirty braccia. From there upward, the plan would be "what shall then be deemed advisable, because in building, what has come before will teach that which must follow."

The machines he invented for hoisting the materials efficiently to their airy destination were truly ingenious. They incorporated the law of the lever, of course, as any machine does, but no deep mathematics was necessary to imagine them. Worthy of Leonardo, these colossal machines must have been seen by the young Leonardo, since the copper ball and cross were not lifted to the top of the lantern by these hoists until 1470. Most impressive of all to Brunelleschi's contemporaries was his method for building the dome without any temporary support from underneath, but here too, mathematics had nothing to contribute.

In spite of the self-sufficiency of the building trade, fifteenth-century architecture was becoming mathematical in a different sense. The rediscovery of the Latin manuscript of Vitruvius's *Ten Books on Architecture* focused attention on the methods of classical architecture and the legendary beauty of Roman buildings. As in every other field of learning, the humanist aim in architecture was to find out the secret of classical excellence. A reading of Vitruvius suggested that one crucial ingredient lay in choosing the right proportions.

Vitruvius was a contemporary of Augustus Caesar, but most of the buildings and ruins that were still visible in the Renaissance dated from the later empire, after the Age of Augustus. And even if Vitruvius could have been a guide to this later flowering of classical architecture, his writing was frustratingly vague and incomplete in many places, offering more of a tantalizing introduction than a key to the secrets of the classical world. Yet while documentary evidence from ancient Rome might have perished, the buildings themselves—the most concrete evidence one could wish for—were still there.

Beginning in the fifteenth century humanist architects began to measure these remains and to tabulate their proportions and dimensions to find the secret, if secret there was. In spite of Brunelleschi's stupendous engineering feat in raising the dome, Manetti, writing some forty years after the fact, seems concerned for Brunelleschi's reputation. He wants to be sure that Brunelleschi is not seen as inferior in the new arena of harmonious proportional design. According to his account, Brunelleschi was the first to go to Rome to make careful measurements of ruins, although there are reasons to doubt this story. For one thing, it was supposedly done in company with the sculptor Donatello, but Donatello was only a teenager, and seems to have been in Florence the whole time. Brunelleschi was a genius, though, and an innovator, and he could have measured Roman ruins if he had wanted to, as the next generation of architects certainly did. All we can really be sure of is that by the time Manetti was writing, the mathematics of proportion was as important as the practical geometry of getting things built at all.

The most influential writer on architecture of the fifteenth century was a near contemporary of Brunelleschi, Leon Battista Alberti. Alberti is also the first to have given a description in writing of the perspective construction. This book, *On Painting*, was written in Latin, but by 1435 it had been translated into Italian, perhaps at Brunelleschi's request, since the Italian edition is dedicated to "Pippo architetto," the affectionate nickname "Pippo" for "Filippo" hinting at a close relationship, although Alberti spent little time in Florence.[2] Alberti was a practicing architect himself, but more than that he was a humanist with a much-admired Latin prose style. His *On the Art of Building*, written around 1450, is consciously modeled on Vitruvius's *On Architecture*, even down to its division into ten books. It mentions modern buildings in only the most cursory way, and doesn't mention Brunelleschi's dome at all. Perhaps Manetti, writing in the 1480s, sensed that in the long term his hero Brunelleschi risked eclipse by the more literate Alberti, and so was motivated to write his biography. A printed edition of the ten books of Alberti appeared in 1485, and the first printed Vitruvius appeared in 1486, but Manetti's biography of Brunelleschi remained a manuscript book until modern times.

Over the next century this print tradition continued, with the appearance of new editions of Vitruvius with commentaries, including a very extensive one by the humanist scholar Daniele Barbaro, and architectural writing by practicing architects in the style of Vitruvius and Alberti, including the influential *Four Books on Architecture* by Andrea Palladio in 1570. Palladio was Barbaro's own architect, having built the Villa Masera for him. Although Palladio is still clearly writing in the manner of Vitruvius, his books are only four, not ten, because he spends much less time on the practical, engineering side of architecture. The discussions that fill many books in both Vitruvius and Alberti: on timber, stone, and brick; on gutters and chimneys; on the orders of columns; on foundations, walls, and roofs; are all crammed into the first book in Palladio. Most of Palladio's books are careful drawings, documenting his researches into Roman buildings. He also includes plans of his own buildings. The carefully labeled dimensions in these drawings are the purest documentary evidence we have of Renaissance architecture, and they make it clear that proportion is all.

9 Figure and Form

A proportion like 4 : 3 in the fifteenth century was not just a mathematical idea. It was musical (the perfect fourth), and it was geometrical. It was associated with the plan of the universe as a whole, with the music of the spheres, and now with the architecture of the ancients. Alberti follows Vitruvius in describing the dimensions of temples and their columns, but he supplements the Vitruvian account with his own observations and measurements of classical ruins. His detailed description of real things seldom follows a simple pattern, but when Alberti suggests his own favored proportions for ideal lengths, widths, and heights, they tend to be the perfect musical proportions. These perfect musical proportions are 2 : 1, 3 : 2, and 4 : 3, and proportions derived from these. In the Neoplatonic milieu of the fifteenth century, he might have done this almost unconsciously, merely making the most self-evident harmonious choices. The classical Vitruvius had introduced yet other significant proportions, the proportions of the human body. As a matter of course the human body, then, came to be associated with music and the universe as well, and all these associations were transferred, quite consciously, to architecture.

Concinnitas

The climax of Alberti's ten books is really the ninth book. (The tenth and last is more of a mopping-up operation, on maintenance and restoration of buildings and sites.) In the ninth book Alberti takes up "a matter which we have promised to deal with all along: every kind of beauty and ornament consists of it; or to put it more clearly, it springs from every rule of beauty . . . For within the form and figure of a building there resides some natural excellence and perfection that excites the mind and is immediately recognized by it."[1] Alberti's name for this intellectually apprehended beauty is *concinnitas.* It is a property to be found everywhere in Nature, he says. "When the mind is reached by way of sight or sound, or any other means, *concinnitas* is instantly recognized." This hint that *concinnitas* is to be found both in sight and in sound is followed by an explicit theory referring to Pythagoras and the musical ratios. Alberti enumerates the Pythagorean harmonies, giving them their customary musical names, and recommends them for use in architectural design: "Architects employ all these numbers in the most convenient manner possible: they use them in pairs, as in laying out a forum, place, or open space, where only two dimensions are considered, width and length; and they use them also in threes, such as in a public sitting room, senate house, hall and so on, when width relates to length, and they want the height to relate harmoniously to both."

As an example of how this theory would be applied, imagine a room that is twice as long as it is wide. This is the ratio 2 : 1, that is to say, the musical octave. Its third dimension, the height, should be chosen harmoniously, as Alberti says above. But how do you divide the octave harmoniously? There are really only two ways. You can go up a fourth 4 : 3 followed by a fifth 3 : 2, or go up a fifth 3 : 2 followed by a fourth 4 : 3. That means the ceiling should be at a height $\frac{4}{3}$ times the width or else $\frac{3}{2}$ times the width of the room. To give actual numbers, the dimensions of the room, width × height × length, could then be 6 × 8 × 12, or else 6 × 9 × 12, in some convenient units. The same principle of harmonic relationships could also be used in choosing dimensions for the subdivisions of a floor plan, and in any other situation where one desires the unity and harmony of *concinnitas.*

Since *concinnitas* is an abstraction, applying equally well to both music and architecture, we cannot say that Alberti advocates realizing music in the solid forms of buildings. Rather, both music and architecture, when they are done well, express the Neoplatonic harmonies of Nature. But since music is the better established and better understood of the two, being an actual science, it frequently sounds as if Alberti conceives of visual things musically, as when he warns the builder of one of his designs, the church of San Francesco in Rimini, that if he were to change the dimensions of the pilasters, it would "untune all that music."[2] Apparently just changing one of the elements in a complex design would be the equivalent of a musical sour note, in Alberti's mind. The *concinnitas* would be upset or spoiled.

Over a hundred years later, Palladio still refers to the concept of *concinnitas,* although not by name, saying, "just as the proportions of voices are harmonies for the ears, so are the proportions of measurements harmonies for our eyes, delightful in the highest degree, without anyone knowing why, except for those who study to know the reasons of things."[3]

Musical proportions alone would be too restricted a vocabulary for visual design, and Alberti is quite ready to approve other proportions as well, just so long as there is a good, rational reason for them. The height of the wall of a circular temple, for example, up to the vault, should be related to the diameter of the circular floor as 1 : 2, or better 2 : 3, or perhaps even 3 : 4. These are instantly recognizable musical ratios, but Alberti seems to have a particular fondness for a fourth possibility, the ratio 11 : 14.[4] This unexpected suggestion has a purely geometrical justification, not a musical one—it is the ratio of one quarter of the circumference of the circular temple to its diameter, namely $\pi/4$, taking the usual Renaissance approximation $\pi \approx 22/7$.

Alberti's insistence on getting these ratios exactly right is emphasized in two places in his own text, where he instructs copyists not to use numerals, which could be easily misread or miscopied, but to write out the numbers as words. In this case he seems to have violated his own injunction, though, since when the book was finally printed, it came out $11 : 4$, not $11 : 14$, which makes no sense. The simplest explanation is that Alberti used the numeral "14" and the "4" was a misprint.

The concept of *concinnitas* is original with Alberti and is not to be found in Vitruvius, even though Vitruvius does discuss music and harmony, briefly, and with apologies for not knowing much about it. The kind of Neoplatonism that comes so easily to Alberti is to be found in the second-century music theory of Ptolemy and Nichomachus, but Vitruvius wrote much earlier. The music of the spheres, and all that it implies for the cosmic significance of the musical ratios, is not something that Vitruvius seems to know about, although he does know a lot of odd things. His commentator Daniele Barbaro felt that he had to supply, in a rather long commentary on music inserted at this point, what Vitruvius had not said, even while he does not see why Vitruvius is talking about music here in the first place. Vitruvius gives a short summary of the ideas of Aristoxenus, who, as we have seen, is explicitly anti-Platonist, and may even be a rare link to an old scientific Pythagoreanism. His music theory by itself doesn't suggest anything about architecture. Vitruvius's discussion of harmonies is to prepare us for a chapter on resonant vessels that the Greeks supposedly positioned among the seats in their outdoor theaters to improve the acoustics. There is no archaeological evidence for this practice, and it remains an ill-understood but intriguing idea for the architecture of theaters. Daniele Barbaro does not see the point. He says, "It seems to me that Vitruvius could have organized his book better, because he puts in many things before there is a need for them." In this case, the discussion of music precedes the topic of sound. To Barbaro, this seems out of place, because for him the discussion of music should precede the topic of proportion. As we have noticed elsewhere, to the Renaissance mind, music can be almost a code for mathematics, and it is not necessarily about sound.[5]

Perhaps this occurrence of musical theory in Vitruvius, however incidental, was enough for Alberti to develop his aesthetic principle of *concinnitas*, or perhaps in the Neoplatonic fifteenth century he didn't need any justification at all, the notion of cosmic harmony being so clearly understood and assumed. On another point, though, Alberti adopted an aesthetic principle from Vitruvius with gusto: the ideal proportions of the human body.

Vitruvian Man

Vitruvius points out that no architectural project requires greater care than the design of a temple, in which the merits and faults, whatever they are, will last "forever." A temple should have proper symmetry and proportion, and above all it should have a "precise relation between its members, as in the case of those of a well shaped man." One might assume that this is just a vaguely intended simile, but Vitruvius goes on to say precisely what some of those symmetries and proportions are. Perhaps Vitruvius is just showing off his erudition again, this time in the field of anatomy, or perhaps he literally means that the proportions of buildings should be those of the body. Renaissance theorists struggled with this question.

Vitruvius lists a great many proportions determined by the human body. A man's face, for example, from his chin to the top of his forehead, should be one-tenth his height, his forearm should be one-fourth his height, and so on. Vitruvius alleges that if a man is placed on his back, with his navel as center, his outstretched fingers and toes will just describe a circle. Also, his height is the same as the width of his outstretched arms, so that a man defines not only a circle but also a square. These observations are rather startling, since the square, and especially the circle, are geometrical symbols of perfection, and here they seem to be derived from the human body. Artists including Leonardo explored the figure of a man in a circle or square as if there might be some deep significance in it.

Alberti seems at some level to take Vitruvius's prescription quite literally, as for example when he argues that the ideal shape for a temple is a circle. This sounds like taking the proportions of a temple from the human body, just as Vitruvius had said. As for the building as a human body, he typically calls the structural elements of a building, like the columns and beams, its "bones," and the exterior paneling its "skin." He even calls the smaller-scale connectors "muscles" and "ligaments." There should be an even number of columns (bones), he argues, because no animal has an odd number of feet. And there should be an odd number of openings, like windows and doors, because the face has two eyes, two nostrils, two ears, but one mouth, making an odd number in all.

Most important, and this idea is echoed in all Renaissance architectural writings, the parts of a building should be well proportioned to each other, so that any one part gives the proportions of the whole. "Look at Nature's own works," says Alberti. "If someone had one huge foot, or one hand vast and the other tiny, he would look deformed."[6] Consistency in proportion is Nature's own law, proved in its exceptions. "The faults of ornament that must be avoided most of all are the same as those in the works of Nature, anything that is distorted, stunted, excessive, or deformed in any way. For if in Nature they are condemned and thought monstrous, what would be said of the architect who composes the parts in an unseemly manner?"[7]

Copernicus published his great work *De Revolutionibus*, on the heliocentric plan of the universe, in 1543. It is a highly technical and geometrical book of astronomy, but in the dedication to Pope Paul III, Copernicus gives an argument for the sun-centered universe drawn from architecture. Surely the architecture of the universe as a whole should manifest that proportionable quality that we expect even in earthly buildings—this hardly needs to be pointed out. But in the Ptolemaic scheme, which his plan might replace, those proportions are nowhere to be found. What Copernicus says of the Ptolemaics is, "they have not been able to discover or to infer the chief point of all, i.e., the form of the world, and the certain commensurability of its parts. They are in exactly the same fix as someone taking from different places hands, feet, head, and the other limbs—shaped very beautifully but not with reference to one body and without correspondence to one another—so that such parts made up a monster rather than a man."[8]

Copernicus is pointing out a deep difference between the Earth-centered and the Sun-centered universes. Ptolemy describes the motions of the planets projected on the sphere of the sky, that is, in just two dimensions. There is nothing in the Ptolemaic description to determine the third dimension, however—the distance from us to the various planets. Mars might be relatively close to us or it might be very far away in the Ptolemaic scheme, and similarly for the other planets. There was nothing even to determine with certainty the relative order of the planets, although there was a conventional order, the one used by Dante in *Paradiso* as he ascends through the heavens of the planets, one after the other. It is this undetermined feature of the Ptolemaic universe that Copernicus calls "monstrous." In his scheme, on the other hand, all the proportions, in all three dimensions, are determined by observation, so that the universe really has an "architecture."

Copernicus implicitly assumes the Vitruvian idea. Although he is talking about harmonious proportions in a three-dimensional structure, he puts it in the language of the human body.

Military Architecture

Both Vitruvius and Alberti devote chapters to the architecture of fortifications. One might expect the nature of warfare to have changed quite a lot in the 1,400 years that separate these two authors, but in fact Alberti follows Vitruvius very closely.[9] Both of them agree that the ideal plan for a city wall is circular. Polygonal shapes are also possible, but Vitruvius cautions that angles offer shelter to the attackers, a point that Alberti repeats. Vitruvius says that the walls should have round towers built into them, projecting out somewhat, and no more than a bowshot apart, so that if attackers come up to the walls, they will be outflanked and exposed to attack by defenders in the towers. If we allow a square shape, what Vitruvius describes is pretty much our conventional picture of a castle, with round towers at the corners. The rooks at the corners of a chessboard, or in the little formation achieved by the move of "castling," are little Vitruvian towers.

Alberti considers two threats to the wall. On the one hand the wall may be battered by a ram so that it is knocked down, or on the other hand it may be undermined so that it collapses. Defenders in the towers, if they can keep attackers away from the wall, will preserve it. As a kind of insurance, a moat is a defense against both kinds of attacks, since it prevents the approach to the wall. What is remarkable is what Alberti does not mention, and what Vitruvius could not possibly have mentioned: cannon. In the fifteenth century cannon were the real threat to the wall. Alberti is a humanist and a scholar. Sometimes he seems so much the consummate classicist that he might as well be a citizen of ancient Rome. But however much Renaissance humanism sought to re-establish the excellence of the ancient world, some of the ground rules had changed forever.

The significance of cannon became obvious to all Italy, if it had not been before, in 1494, with the invasion of Charles VIII of France, the same campaign that brought down Duke Ludovico Sforza and drove Luca Pacioli and Leonardo da Vinci from Milan. As Charles's army with its siege cannon demonstrated repeatedly, the walls of any fortress could be breached within hours, without anyone even having to approach the wall. The whole theory of fortification, which until now had favored the defenders, suddenly needed revision, and classical literature was not going to help.

What emerged over the next decades was a new design that stipulated that forts should be polygonal, typically regular pentagons, with angled bastions at each corner. These bastions had a very definite shape and projected out from the pentagonal ground plan like the round towers in the Vitruvian plan. Such forts were constructed all over Italy, and indeed all over the world. Many eighteenth- and nineteenth-century examples can be found even in the United States. Fort McHenry in Baltimore Harbor, the inspiration for "The Star Spangled Banner," is a pentagonal fort, and Fort Warren in Boston Harbor is another, as is Fort Sumter in Charleston, South Carolina, where the American Civil War began. Nor should we forget the U.S. Department of Defense headquarters in Arlington, Virginia, the biggest pentagon of all, although its shape is more symbolic than functional.

It is natural to wonder whether this adoption of the pentagonal shape was influenced by the occult significance of the pentagon, the properties rhapsodized by Luca Pacioli in *De Divina Proportione*. Pacioli, after fleeing the French at Milan, had turned up in Venice, where he wrote an architectural section for *De Divina Proportione*, cribbing freely from Vitruvius on the proportions of the human body and the design of columns. In explaining which subjects he would treat and which he would not, he remarks on what he had experienced firsthand. He notes that the subject of fortification had changed considerably "because of the new artillery, that didn't exist in the time of Vitruvius, so that we will leave this subject for now, and keep it for more extensive discussion on another occasion." Pacioli never contributed to the theory of fortification, but he did perceive the problem. He never suggested a pentagonal fort. Oddly enough, in his architectural writing Pacioli never suggests using the golden ratio in a building or in any other real object. Its associations for him are completely abstract. This is a reminder that philosophical mathematics has literally no application, and is simply not an applicable subject.

If the pentagonal fortification did not emerge from the abstractions of philosophical mathematicians like Pacioli, where did it come from? A pentagon encodes the golden ratio in its dimensions. Perhaps someone more imaginative had suggested that the magic of the golden ratio might lend a special strength to a pentagonal fort? The truth is much more prosaic. Military architects were practical people, with their feet on the ground. Their arguments for the new design came from experience and from purposeful theory.

The theorists of fortifications actually used some mathematics in laying out their ground plans. Their program emerged out of a series of conferences convened by Pope Paul III on the defense of Rome.[10] The new theory conceded the inevitable, that cannon would breach the walls no matter how thick they were. The new pentagonal design was intended to neutralize this fact by making it useless for the attacker to do so. That was the role of the bastions at the corners. The bastions were like the towers in the Vitruvian design, but lower, and equipped with their own cannon to rake the wall with enfilading fire, making a rush into the breach suicidal for the attackers. For this purpose the walls between the bastions had to be straight, not circular, and about the length of a cannon shot (so that the whole wall was defended). Thus the fortification would consist of straight segments, making it a polygon. There were reasons, having to do with the shapes of the bastions, why the triangle and the square were not good choices for the overall polygonal shape. The angles in the plan had to be more obtuse. Dividing the desired circumference of the fort by the length of one side (which might be as much as the distance of a cannon shot) gave the number of sides needed, but since the length of a side could be rather long, the number of sides was sure to be rather small. And since three and four were ruled out, it essentially always came out five.

Adapting these ideas to the complex topography of a real site, where the ideal plan couldn't simply be plopped down unchanged, might require some geometrical inventiveness. It would also require experience of warfare and the imagination to know what use an enemy could make of the peculiar features of a place. Such expertise was rather distant from the concerns of civilian architects, and military architecture soon became its own specialty, a highly respectable profession.

Galileo's first patron was such a military architect, Guidobaldo del Monte, Inspector of Fortifications for Tuscany. He was a marquis and a high-ranking member of the Medici court, when he chose to be in attendance. Normally he lived on his estate at Montebaroccio, conferred on his father by Guidobaldo II, Duke of Urbino. Guidobaldo del Monte had a first-rate mathematical education, and when he saw Galileo's first mathematical efforts, he was deeply impressed. His mentoring of the young Galileo made Galileo's career possible.

10 The Dimensions of Hell

At the age of twelve Galileo became a novice of the order of Camaldolese monks at the monastery in Vallombrosa, where he had been sent to school. Ultimately he did not, of course, become a monk—his father put a stop to that. But something vaguely similar happened when he went to the university in Pisa to study medicine. In 1583, at the age of nineteen, under the influence of the geometer Ostilio Ricci, Galileo made what must have seemed to his father like another hopelessly otherworldly career choice: he decided to abandon medicine and devote himself to mathematics. In 1585 he dropped out of the university.

Galileo's prospects for a career in mathematics were slim. He tried teaching mathematics publicly in Siena for a short time in 1586, but his family aimed to be more patrician than this, even if they didn't have money. What he really needed was a position at a court or in a university, but he had almost no qualifications. He was essentially self-taught, and anyway, such positions hardly existed. In 1587 one of the two chairs in mathematics at Bologna opened up and Galileo applied for it, but it went to Giovanni Antonio Magini.

It is in this period that Galileo and his father conducted the experiments on music and stretched strings. Galileo, working on his own, also proved some nice results in the style of Archimedes on centers of gravity in parabolic figures. He was very happy with these theorems, and in the sense of pure mathematics they are perhaps his most beautiful work. Over fifty years later he published them as an appendix to *Two New Sciences* even though, as he says, they had long since been superseded by results of others. At the time he sent them to the most prominent mathematicians in Italy, including Christopher Clavius in Rome, who had just overseen the Gregorian calendar reform, and Guidobaldo del Monte, the Inspector of Fortifications for Tuscany.

Guidobaldo admired the theorems and adopted the young Galileo as a kind of protégé. Guidobaldo himself came from an old aristocratic family. He had an independent income and could afford to indulge an interest in mathematics without making it pay. He was well-connected at court and more than happy to correspond with Galileo about mathematical questions, but he still didn't have the clout to arrange something that would secure Galileo a living.

All this changed with the sudden, strange, and unexpected death of the Grand Duke Francesco de Medici in 1587. Francesco I and his second wife, formerly his mistress, Bianca Capello, had gone to a Medici villa outside the city to enjoy hunting, and there they died within twenty-four hours of each other, of apparently unrelated causes. An inquest turned up no evidence of foul play. Francesco had suffered severe stomach pains, but any attempt to demonstrate that he had been poisoned would have been complicated by his interest in alchemy and his willingness to concoct his own medicines. He might have poisoned himself. The most obvious beneficiary of his death was the new grand duke, Francesco's brother Ferdinand, but there could be no reason to suspect him of complicity. He had been living the comfortable life of a cardinal in Rome, a life that seemed to suit him very well. He had to give up the cardinalate to become Grand Duke Ferdinand I. Now, in a stroke of fantastic luck for Galileo, his cardinal's hat went to Guidobaldo del Monte's younger brother Francesco, who became the new Medici cardinal. Suddenly Galileo had, in Guidobaldo's brother Francesco, a much more powerful patron. Within a few months the university dropout would be back at Pisa, this time as professor.

Events had moved rather fast, and the new mathematician, although he was the son of a well-respected father, was virtually unknown. It did not really help that he had proved some arcane Archimedean results that not one person in a thousand would understand. It was therefore arranged that he should lecture to the Florentine Academy, a very prestigious forum, on some aspect of mathematics that might appeal to a more general audience. Galileo rose to the challenge. He prepared two lectures on the architecture of Dante's Inferno.[1]

The Inferno Lectures

The Florentine Academy was not a dispassionate learned society. It had been established in 1540 by Cosimo I de Medici as part of a conscious strategy of glorification of the ruling family and the city of Florence.[2] The Medici found this advisable because their legitimacy as hereditary rulers was not so self-evident. Although the Medici had controlled Florence through much of the last century and more, they had done so within the nominal structures of a republic. Cosimo I was not of noble birth. He only became duke in 1537, having seized unchallenged control of the city by violence, bringing the venerable Republic of Florence to an end, and he only became grand duke by edict of Pope Pius V in 1569, that is, in Galileo's own lifetime. The Medici were in control, but they needed validation. The Florentine Academy was a Medici institution engaged in developing a mythology of the city in which Medici rule was a natural and fortunate outcome of its entire history. Galileo was being given the chance to show what he, as a

Medici intellectual, could contribute to the glory of Florence and its rulers.

A bare summary of the Inferno Lectures scarcely hints at the drama under the surface.[3] Superficially, Galileo seems to consider two architectural models of Dante's Inferno in minute detail. He gives all the relevant dimensions, with a lot of tedious arithmetic to keep track of how things fit together, citing evidence from Dante for why this or that number has been chosen, and ultimately he decides that one of the models is the better one. This does not sound as if it would even keep his audience awake, yet the lectures were an unqualified success, and were apparently discussed for years afterward. A letter from one Luigi Alamanni summarized Galileo's argument accurately more than five years after the event, even though he had been unable to find a written copy of the lectures and was relying on memory.[4]

Florentines revered their Dante, who, despite his exile and his railing against the injustice done to him by his native city, was still an illustrious son of Florence, the one who had elevated the Tuscan language to the undisputed standard of literary Italian. Just contributing to the appreciation of Dante was already a very safe strategy for a speaker. Florentines knew *The Divine Comedy* intimately, and not just the poem, but also the many commentaries on it, a popular literary form in Florence and throughout Italy. In his lectures Galileo took one of the architectural models for the Inferno from the commentary of none other than Antonio Manetti, Florence's fondly remembered "citizen and architect," the biographer of Filippo Brunelleschi. The other architectural model was the design of Alessandro Vellutello of Lucca.

No one in the audience would have to be told what this meant. Lucca had been the site of a Florentine humiliation in the year 1430. A Florentine army had been besieging Lucca without success when Filippo Brunelleschi himself, this time in the capacity of a military engineer, conceived the idea of diverting the river Serchio so as to isolate Lucca within a lake and compel its surrender. This grandiose plan went catastrophically awry when a dike failed and the Serchio rolled over the Florentine army camp instead. Now that battle was about to be fought again, with Manetti and Vellutello standing in as intellectual proxies for Florence and Lucca, and there was only one possible outcome. Vellutello was going to get creamed. The only question his audience might have had for Galileo was, how was he going to do it? One can just picture the Academicians licking their lips as they sat in their chairs. Young Galileo begins in a way that seems even-handed, but as the second lecture proceeds, he becomes more and more sarcastic about Vellutello's plan until in the end he seems to be defending the virtuous Florentine Manetti against the stupid and thoughtless Vellutello.

Manetti's Inferno is a cone-shaped region in the earth, with its vertex at the center and its base on the surface, centered on Jerusalem. Since Galileo is a master of exposition, let him describe it:

> Imagine a straight line which comes from the center of the earth (which is also the center of heaviness and of the Universe) to Jerusalem, and an arc which extends from Jerusalem over the surface of the water and the earth together to a twelfth part of its greatest circumference: such an arc will terminate with one of its extremities on Jerusalem; from the other let a second straight line be drawn to the center of the earth, and we will have a sector of a circle, contained by the two lines which come from the center and the said arc; let us imagine, then, that the line which joins Jerusalem to the center staying fixed, the other line and the arc should be moved in a circle, and that in such motion it should go cutting the earth, and move itself until it returns to where it started. There will be cut from the earth a part like a cone; which, if we imagine it to be taken out of the earth, there will remain, in the place where it was, a hole in the form of a conical surface, and this is the Inferno.[5]

As an aside to the mathematically adept, Galileo gives the volume of this region, which he was able to compute from his study of Archimedes, as less than $1/14$ the volume of the entire Earth.[6]

The various levels of Manetti's Inferno are regularly spaced, for the most part, with $1/8$ the radius of the Earth between each level and the next. In particular the first level, Limbo, is at a depth of $1/8$ the radius of the Earth below the surface, and the shell of material down to this depth forms a cap of this thickness, roughly 405 miles in Galileo's estimation, over the conical cavity below. Vellutello's Inferno, by contrast, is much smaller, located near the center of the Earth, and only about $1/10$ the radius of the Earth in height, making it, as Galileo is quick to say, ridiculously small, only $1/1,000$ the volume of Manetti's.

In putting it this way, Galileo used a rhetorical trick that he was very fond of. It sounds much more impressive to say that the smaller model is only $1/1,000$ the size of the larger one (in volume) than to say that it is only $1/10$ the size of the larger one (in height), even though these are exactly the same statements about geometrically similar figures. Whatever the scale factor in length, you must cube it to get the scale factor in volume. By the same reasoning, the scale factor in area is the square of the factor in length, so that the "ceiling" of Vellutello's Inferno, for example, being an area, would be only $1/100$ the ceiling of Manetti's. Galileo tended to describe contrasts in size in terms of volume, if he could, rather than length, just so that he could cube the scale factor, making the contrast sound more impressive. When he describes his best telescope later on, for example, in the *Starry Messenger*, he says that it magnifies lengths by a factor of more than 30, but it magnifies areas by a factor of almost 1,000 (the square of 30 would be 900). It doesn't make sense to consider volumes here, because telescopes see areas, not volumes, but Galileo actually does say that his telescope makes the volume look 27,000 times bigger. The scaling behavior of areas and volumes in geometrically similar figures was about to play a most unexpected role in the story.

Part of the emerging ridicule of Vellutello's scheme is that his plan is so small. Manetti's scheme starts from the Earth as a whole, and the entire Earth sets the scale. But Vellutello's scheme starts from Lucifer, fixed in the ice regions at the center of the Earth, and Lucifer's proportions set the scale. As Galileo puts it, emphasizing the diminutive size,

> we should understand a pit which has a diameter of one mile, both at its top and at its bottom, and one mile is also its depth, and at its bottom there is ice, in the shape of an enormous grindstone (we are supposed to believe this) 750 braccia thick. And this ice should be divided into 4 circles, so that one surrounds the next, and in the middle of the smallest there should be a little pit, again in the shape of a grindstone, whose depth is the thickness of the ice, that is, 750 braccia, and in the middle of this is the center of the world, and Lucifer would be standing in this little pit.[7]

Starting at the center of the Earth and deriving the dimensions of the Inferno from the dimensions of Lucifer actually makes a certain amount of sense. There is a direct invitation in Dante to reconstruct the size of the ice regions from information that he provides, and both Manetti and Vellutello take up this invitation. The lines in Dante are:

> The emperor of the realm of grief [Lucifer] protruded
> From mid-breast up above the surrounding ice.
> A giant's height and mine would have provided
> Closer comparison than would the size
> Of his arm and a giant. Envision the whole
> That is proportionate to parts like these.
>
> *Inferno* XXXIV, 28–33 (tr. Pinsky)

Galileo leads us through the argument. First of all, Lucifer's navel is at the center of the world, like a kind of perverted Vitruvian man. Since the ice extends up to his mid-breast, one quarter of his height, we would know the radius of the ice sphere if we knew Lucifer's size. But Dante actually tells us Lucifer's size, indirectly: Lucifer's arm to a giant is greater than a giant is to Dante. We know that Dante was of average height, says Galileo, that is, three braccia. Therefore we have only to find the size of a giant, and Dante tells us this too. From *Inferno* XXXI, 58–60, the giant Nimrod's face is the height of the Pinecone at Rome. (This bronze sculpture was originally an ornament of Hadrian's tomb. Later it was exhibited outside St. Peter's Cathedral, where it suffered some damage, and what is left of it can still be seen at the Vatican.) A normal man is eight heads high, according to Galileo, and therefore the height of a giant is eight times the height of the Pinecone, and the problem is solved. Vellutello had taken slightly different figures, for example that the giant would be nine heads high, a proportion that can be found in the writings of Albrecht Dürer,[8] but Galileo claims that such proportions are very unusual, and that eight is more likely. Needless to say, Manetti had used eight, not nine, and arrived at a height of 2,000 braccia for Lucifer where Vellutello found 3,000 braccia. One fourth of this, the proportion of a man between navel and mid-breast, is the thickness of the ice, and thus 750 braccia in Vellutello's scheme.

Manetti's scheme describes, overall, a conical Inferno with broad terraces determined by a geometrical construction that applies to the whole structure, with a simple scale factor relating each level to the next, but the Vellutello scheme, building from the bottom up, starting with Lucifer, does not really have a module that is suitable for repetition. The first several levels are built from cylinders, progressively larger as one goes up from the center, but then, for the upper levels, the scheme switches to conical units.

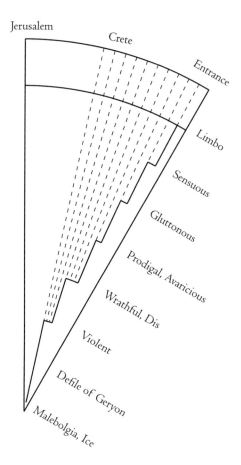

Jerusalem
Crete
Entrance
Limbo
Sensuous
Gluttonous
Prodigal, Avaricious
Wrathful, Dis
Violent
Defile of Geryon
Malebolgia, Ice

Manetti's Inferno occupies a huge conical volume inside the Earth, obtained by rotating this figure about its left edge. It is covered by a roof 400 miles thick.

Galileo criticizes this conception on architectural grounds, quite apart from what evidence Dante might give for it: "And first, if we will consider the one and the other design without having regard to any place in Dante, or any argument that would persuade us that the one or the other is more plausible or believable in representing Dante's intentions, but only contemplating the disposition of the whole and of the parts, and in sum, so to speak, the architecture of the one and of the

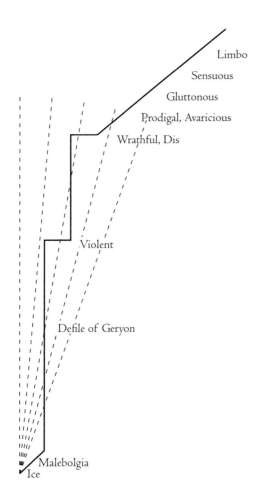

Limbo

Sensuous

Gluttonous

Prodigal, Avaricious

Wrathful, Dis

Violent

Defile of Geryon

Malebolgia
Ice

Vellutello's Inferno is made of cones and cylinders, obtained by rotating this figure about its left edge, and occupies a small volume near the center of the Earth. The vertical direction (toward the center of the Earth) is indicated by dashed lines, illustrating Galileo's contention that the structure would collapse inward of its own weight.

other, we will see, as it seems to me, how much better is the design of Manetti, composed of parts more similar to each other."[9]

Now Galileo points out a rather serious flaw in the lower, cylindrical regions of Vellutello, the pits that have the same diameter at the bottom as at the top. Such pits actually have walls that are unsupported below, because in effect they are overhangs. The direction "down" is toward the center of the Earth, but if you start at the upper rim of one of these cylinders and draw a line to the center of the Earth, it does not go down the side of the cylinder, as it should do if the upper part is supported by the lower part. Rather the line to the center goes through the empty interior of the cylinder, so that the upper part is not supported by anything and will surely collapse. The descents in Manetti, on the other hand, are all vertical descents, in the sense of being directly toward the center of the Earth along radial lines.

It is not enough just to demolish the lower part of Vellutello's model. Galileo will also demolish the upper part. There the descents in Vellutello are again not vertical, but instead of being overhangs, exceeding the vertical, they are more gentle than vertical, like mountainsides. This might seem plausible, Galileo says, but the evidence of the poem is against it, because if the descents were like mountainsides, one could descend anywhere, but Dante in his journey is only able to descend at certain very special places, guarded in each case by an appropriate demon. At other places, therefore, the descent must be vertical, and hence impossible, as in the model of Manetti. "And I believe it is like this so that the sinners of the lower levels, where the torments are greater, cannot escape and flee to the higher levels, to lesser torments; and it seems that this is what Dante meant when he put, at each place where one could go from one level to another, a demon."[10]

On the question of overall size, Galileo answers a criticism that had apparently been made of Manetti's model, that the roof over the great conical cavity would not be strong enough to support itself, and would collapse into the pit:

> Here one might oppose that the Inferno cannot be so large as Manetti makes it, since, as some have suspected, it doesn't seem possible that the vault that covers the Inferno could support itself and not fall into the hole, being so thin, as is necessary if the Inferno comes up so high. And especially, beyond being no

thicker than the eighth part of the radius of the earth, which is 405 miles more or less, some of it must be removed for the space of the Grotto of the Uncommitted [not part of the Inferno proper], and even more must be removed [on the top] for the very great depth of the sea. To this one answers easily that such a thickness is more than sufficient; for taking a little vault which will have an arch of 30 braccias, it will need a thickness of about 4 braccias, which not only is enough, but even if you used just I braccia to make an arch of 30 braccias, and perhaps just $\frac{1}{2}$, and not 4, it would be enough to support itself; and knowing that the depth of the sea is a very few miles, or better, even less than one mile, if we believe the most expert sailors, and assigning as many miles as seem necessary for the Grotto of the Uncommitted, a determinate measure not being given by the Poet, if this together with the depth of the sea comes to 100 miles, the said vault will still be very thick, and far more than is necessary to hold itself up.[11]

Galileo plays with the material, introducing arguments of many kinds, but all tending the same direction. Sometimes it is just repeating an argument previously made, but with some sarcasm: "The novel opinion of Vellutello that Lucifer would be 3000 braccia high and not 2000, as Manetti would have it, originating from nothing other than wanting to measure the Pinecone before it was broken, and wanting the giants to be 9 heads high, does not seem very believable to us. On the contrary, it is believable that Dante, if he even measured it, measured the Pinecone as it was in his time, and that he believed the giants to be of ordinary proportion, and not of that rare proportion which would make them 9 heads high."[12] Or, getting near the end now, he questions Vellutello's competence: "Similarly, there is neither argument nor authority that persuades us to believe that the ice regions would be like grindstones, and not like spheres. On the contrary, since the Poet himself called them spheres in the last canto, it is not without temerity to want to say that they have the shape of grindstones, as if a genius such as Dante was would have lacked the words to express his own conception."[13]

On the very basic question of whether Dante circled around to the right (as Manetti thought) or to the left (as Vellutello thought), Galileo quotes Dante:

> As you know well,
> The place is round; although you have come far
> Always to the left descending down to the pit.
>
> *Inferno* XIV, 124–126 (tr. adapted from Pinsky)

This sounds as if Dante has been circling always to the left, just the opposite of what Manetti had said. How to make it say right, and not left? Galileo had a way of starting with evidence that seemed to favor his opponents, then turning it around so that it proved the opposite, a trick relished and admired by his friends. In this case he says,

> In these verses, if we put the words "always to the left" together with those before, saying "although you have come far always to the left," making a pause in the middle of the line, they support the opinion of Vellutello; but if we make the pause at the end of the second line, joining the words "always to the left" with the following ones, in this way: "always to the left descending down to the pit," they favor the opinion of Manetti. Now in an uncertain interpretation, who would not find it better to make the pause at the end, rather than in the middle of the verse?[14]

Whether the argument is right or wrong is largely immaterial if it is self-confident, entertaining, and witty, as it certainly is. In a final flourish of *sprezzatura,* seeming not to be too serious and to have done it all effortlessly, Galileo finishes:

> But since proceeding either to the right or to the left is not very important to our principal intention, which has been to set out the location and shape of Dante's Inferno, and at the same time to defend the ingenious Manetti against the false calumnies that he has unjustly received on this subject, and especially since they have stung not just him alone, but the whole most learned Florentine Academy, to which for many reasons I feel myself most obligated; and having showed, with what little ingenuity I possess, how much subtler is the invention of Manetti, I bring my argument to a close.[15]

It had been a boffo performance, and he got the professorship.

Disaster

Not long after the lectures—we will never know exactly when, because he kept it a secret—Galileo realized that he had made a catastrophic mistake. Manetti's model of the Inferno was impossible. The argument that Galileo had made about the roof being strong enough to hold itself up, based on a scale model, was not just wrong, it actually proved the opposite: that the roof would instantly collapse.

Fifty years later Galileo finally published a viable scaling theory in his last book, *Two New Sciences,* but he did it without referring to the Inferno Lectures at all. Quite the contrary, he invented an entirely different context for the problem. *Two New Sciences* opens in the Venetian Arsenal, where Salviati, Sagredo, and Simplicio, the dialogue's participants, are admiring the great ships under construction. The old foreman has been telling them that the larger ships require more scaffolding and bracing than the smaller ones, because they are more in danger of splitting under their own heavy weight. Sagredo says he can't believe this. Surely geometry governs the situation, and the propositions of geometry have nothing to do with scale: the properties of a triangle are the same whether the triangle is large or small. As Salviati begins to enlighten him, and to prove that scaling is more complicated than this, and that structures are weaker in a certain precise sense when they are scaled up, Sagredo responds in oddly emotional language: "My brain . . . reels. My mind, like a cloud momentarily illuminated by a lightning-flash, is for an instant filled with an unusual light, which now beckons to me and which now suddenly mingles and obscures strange, crude ideas."[16] The language seems wildly out of proportion to the topic under discussion, but it is apt if it records Galileo's own feelings of panic as he realized the mistake he had made in the Inferno Lectures, which was on precisely this point, the relative weakness of structures that are scaled up. Galileo had described a roof like a dome thirty braccia wide and four braccia thick (not so different from Brunelleschi's dome, in fact) as evidence that a roof 3,000 miles wide and 400 miles thick would be strong enough. The small scale model *would* be strong enough, but just barely—that Brunelleschi had succeeded in building a dome of a similar size was almost a miracle. Scaling it up by one braccia to 100 miles is a factor of 300,000. As Galileo had now realized, the scaled-up dome of Manetti's Inferno would be weaker by that same enormous factor and would instantly fall into the pit.

There is nothing so terribly wrong with making a mistake in a scientific or mathematical problem. Quite the contrary, a mistake may be a golden opportunity to understand a problem better in the purely scientific sense. Galileo developed a new theory of scaling in coming to understand his mistake. In our scientific culture he would have published his new insight immediately, and gotten a lot of credit for it. But the Inferno Lectures were never scientific. Rather, they were an intellectual entertainment with strong political overtones, and it was essential that the Florentine side win. In bringing in the argument of the scale model, Galileo had inadvertently furnished his next rival with the tool to humiliate him, and the Florentine Academy along with him. It was worse than a disaster, it was like professional death. Galileo had to prepare for the encounter with that challenger.

We don't know when Galileo worked out the scaling theory, but there are hints that it was very soon after the Inferno Lectures. Sagredo's emotional response makes less sense if it describes a realization many years after the event. Had the mistake gone unnoticed for years, and had Galileo only then realized it, he would have reacted more calmly. The Venetian setting of *Two New Sciences* is another hint. Galileo held his first professorship in Pisa for only three years, 1589–1592, and then obtained the chair in Padua, in the Venetian Republic, very convenient to Venice itself. Galileo was a favorite among the Venetian intelligentsia, who included the historical Giovanni Francesco Sagredo of the dialogue. In his first years there he even consulted for the Venetian Arsenal concerning the placement of oars on large ships. If we take Salviati to be Galileo's own spokesman in the dialogue, as he so often is, then the opening scene seems to say that Galileo already understood the scaling theory before he saw the shipbuilding, and that he was already prepared to explain it to his friends.

Indeed, Galileo must have worked on explaining this new idea in an entertaining and compelling way. He had been wrong about Manetti's Inferno, but he didn't have to leave the subject there, if he could somehow turn the tables and take the argument in a new direction, delighting his courtly audience, if it came to that, with some unexpected novelty and yet more audacious ingenuity. In his prepared defense he first developed a thorough understanding of what exactly goes wrong when you try to scale things up. Then he applied the theory to some completely different problems, like ships, and waited for the time when he would have to use it.

Naturally enough, Galileo seems to have regarded the Inferno Lectures from that point on as an embarrassment. His first biographer, Viviani, was unaware of the lectures, despite living in Galileo's house during his last years, collecting Galileo's stories, and devoting himself to Galileo's memory after his death. Because Viviani is the source of most stories about Galileo, subsequent biographies have also said little or nothing about the Inferno Lectures, which were only rediscovered in the nineteenth century. They seemed so uncharacteristic of Galileo, as he was imagined at the time, that their authenticity was initially questioned. Thus one can plausibly conclude that Galileo never told Viviani about them. This circumstance is rather odd when you think that they were very likely the pivotal opportunity that launched his career. The 1594 letter of Luigi Alamanni already alluded to indicates that Galileo was unhelpful to someone who wanted to get a copy of them.[17] The letter says, "About that lecture of Galileo, he is in Padua, and I have not been able to get it from him." There were copies with Bacio Valori in Florence, who had been consul of the Florentine Academy at the time, including one written out in Galileo's own hand, but Galileo apparently did not volunteer this information.[18]

In the end, the challenge never came. Apparently no one noticed Galileo's error, and he finally published this material on scaling fifty years later, as the first of the *Two New Sciences*. He must have lived all his life with the knowledge that he was vulnerable because of this early and very public mistake, but the threat became gradually less as the years went by and no one noticed. His move to the University of Padua in the Republic of Venice, in 1592, made him somewhat more secure, as he was no longer so immediately dependent on the Medici (although he was still their subject). By the time he moved back to Florence in 1610 as court philosopher to the grand duke, he must have felt quite confident. And by the time he published the scaling theory in *Two New Sciences*, there may have been no one left alive who had been in the audience and heard the original lectures, so that there was no compelling reason to recall Manetti's Inferno. After all, it would still have been an embarrassment to the Florentine Academy. In the end, despite any potential awkwardness, Galileo seems eager to bring the scaling theory to a long-postponed resolution, losing no time in getting right to the subject in the opening pages of his last book.

The basic idea of Galileo's scaling argument, as he finally develops it in *Two New Sciences*, goes back to the behavior of volumes and areas under scaling. If an object is scaled up by the factor 10, for example, then its volume scales up by the factor 1,000, and therefore its weight also increases by a factor 1,000. What prevents it from breaking are the mysterious adhesions that hold things together along any surface that represents a potential break. But such a surface, being an area, only scales up by the factor 100, not 1,000, and it makes sense that the number, and hence strength, of these adhesions would also scale up by only the factor 100. Thus the weight grows more than the strength, by a factor 10 in this example, so relatively speaking the strength grows less than the weight that it has to support, by a factor of 10. Galileo concludes that anything, no matter how strong, will be too weak to support its own weight if it is scaled up sufficiently.

One might have expected this scaling phenomenon in architecture to be common knowledge, but apparently it was not. Galileo certainly didn't know it at the time that he gave the lectures, and no one in the audience seems to have known it either. Architects routinely made models of their designs as part of the process of getting all the proportions right, as Alberti recommends, and builders often worked from these models, but no one seems to have suggested that a model might actually be impossible to build at a larger scale. The enormous dome of Santa Maria del Fiore was required by a design made decades before, and it turned out to be just barely possible to build it. As cities vied to build ever more impressive cathedrals, however, the size limit was being reached, and there were cases, like the cathedral at Beauvais, of catastrophic collapse.

Curiously, in *Two New Sciences*, Galileo never applies his scaling theory in any detail to architecture, mentioning only once in passing that buildings cannot be scaled up indefinitely. This seems like the most important practical consequence of the theory, but perhaps this application was too reminiscent of the collapse of Manetti's gigantic Inferno, and if this material was really part of a defense against a challenger, he would have wanted to deflect attention from that architectural debacle. The application that Galileo does address, charmingly, is the scaling up of animals. As animals get larger, they get proportionately weaker, according to this theory, at least if larger animals are simply scale models of smaller animals. To some extent, this does seem to be true: "a small dog could probably carry on his back two or three dogs of his own size; but I believe that a horse could not carry even one of his own size."[19] And if larger animals are not to be weaker, then something else would have to change. Galileo imagines that their bones could become proportionately thicker to support more weight.

> It would be impossible to build up the bony structures of men, horses, or other animals so as to hold together and perform their normal functions if these animals were to be increased enormously in height; for this increase in height can be accomplished only by employing a material which is harder and stronger than usual, or by enlarging the size of the bones, thus changing their shape until the form and the appearance of the animals suggest a monstrosity. This is perhaps what our wise Poet had in mind, when he says, in describing a huge giant:

> You cannot reckon his height
> So misshapen is his bulk.

The quotation is from Ariosto's *Orlando Furioso*.[20] The appearance of a poetic giant in *Two New Sciences* reinforces the suggestion that this material was originally part of a defense against a challenger to the Inferno

Lectures. Galileo realizes by now that giants can be neither eight heads high nor nine heads high, and he even has a wise Poet to back him up, although it is not Dante this time but Ariosto. A courtly audience might not follow the scaling analysis, but they would applaud this poetic trick just as they had applauded the original lecture. It doesn't seem to matter that *Two New Sciences* addresses a vastly different audience. The original material stays in, even though Ariosto's giant seems peculiarly out of place in one of the founding documents of modern physics.

The Inferno Lectures had turned into something unexpectedly deep, as the purely geometrical idea of scaling became an insight into the strength of natural and artificial structures. The efficacy of old mathematics interpreted in this new way was unexpected even to Galileo himself. It was not the function of Renaissance mathematics to produce new insights into things. Rather mathematics itself was the insight, to be studied for itself, or in the service of astronomy or music. How had the rules suddenly changed?

A Second Look at the Inferno Lectures

Galileo's lectures were in support of his candidacy for the mathematics professorship, and thus they should have addressed astronomy, the subject that he would be teaching, after all. But Galileo didn't know anything about astronomy, and had never studied it or shown any interest in it. As it turns out, though, the Inferno Lectures do start with astronomy, unlike any previous Inferno commentary. The opening sentence is:

> If it is an amazing and wonderful thing that men have been able,
> by long observations, unceasing vigils, and perilous explorations,
> to determine the measure of the heavens, their motions fast and
> slow, their proportions, the sizes of the stars—not just the

nearby ones but the distant ones as well—and the geography of the land and of the seas: things which, either in whole or in large part, are sensible to us; how much more marvelous must we esteem the investigation and description of the location and form of the Inferno, which, buried in the bowels of the earth, hidden from every sense, is known to no one by experience; where, although it is easy to descend there, it is nonetheless very difficult to return, as our Poet tells us so well.

The transition to the Inferno is accomplished rapidly, to be sure, but the nod to astronomy is there.

It is worth noticing how peculiar Galileo's earliest characterization of astronomy is. The proportions of the heavens, and especially the sizes of the stars, are not things that standard astronomy had determined or even considered.[21] Perhaps, though, Galileo is setting up an analogy that gives his real subject a mathematical heft that it had not had before. Proportions and sizes don't have much to do with astronomy, but they are exactly what he is about to deal with in the Inferno. He seems to be representing the Inferno as a subject for geometry on a par with the heavens.

Galileo's lectures were in a tradition of Dante commentary that began with Cristoforo Landino, whose sumptuous 1481 edition of *The Divine Comedy* included figures by Sandro Botticelli and first introduced Manetti's ideas about the Inferno. Those ideas are rather sketchy in Landino's version, and a near-contemporary drawing by Botticelli of the plan of the Inferno does not even show vertical drops from one level to the next, that feature which Galileo considered so important. The model evolved over the succeeding decades, reaching nearly finished form in a little book of Pierfrancesco Giambullari (1544) about the same time that it was attacked by Vellutello.

All Dante commentaries aim to help the reader understand the text's obscure meanings, and these Inferno commentaries in particular aim to tell us as precisely as possible where Dante and Virgil are located in their subterranean journey, a question that probably would not occur to a modern reader. Manetti and his disciples are not describing a real Inferno and a real journey: they are all explicit about this. Rather they aim to reveal the poet's *mind*, his intention. Landino compares Dante's Inferno with the underworlds of the *Odyssey* and the *Aeneid*, and marvels at how, in contrast to those poetic predecessors, Dante "by his most subtle mind and by the discipline of mathematics" shows it to us almost as though it were there before our eyes. They are perfectly serious about wanting to understand and elucidate Dante's conception, his thought. In fact, it made more sense in the Renaissance context to apply mathematics to Dante's poem than it did to apply mathematics to the accidental world of our experience, since Dante's world clearly was mathematical, as ours presumably is not.

Galileo, too, although he speaks ironically when he is defending Florence against her commentator detractors, is not at all ironic about the commentary enterprise itself. He even adds his own carefully considered improvements, although he ascribes them to Manetti. He says, for example, after describing the depth and structure of Malebolgia, the eight circle, "We have added this discourse, and the demonstration of the distance from Malebolgia to the center, to that which has been written by the friends of Manetti as an explication of his discoveries, it seeming to us, as indeed it is, that they had neglected to explain the most subtle inventions of Manetti's noble genius." That is, Galileo worked on this problem seriously, just as his predecessors had done.

In contrast to his predecessors, though, Galileo takes a fundamentally geometrical, not arithmetical, point of view. He begins with a geometrical theorem about proportions: sectors of circular arcs subtending the same angle are in the same proportion as their radii. This gives him a flexible framework for considering the design. Each part of the architecture is a model for each other part, as Galileo explicitly says, and he can mentally move them about and adjust them. He was willing to assign numbers advisedly, so as to simplify the arithmetic.[22] The resulting structure is not inconsistent with the Manetti school of thought, but it is less accidental, less arithmetical, and more like astronomy in having a fundamental aesthetic to which the data are adapted.

Galileo's conception of Manetti's plan emphasized the repeated elements, spherical sectors, an extended analogy relating each part to each other part. When Galileo faced the problem of the roof, there was nothing else in the model that it was analogous to—the Inferno has only one roof—so he went outside the model to find an analogy, a physical dome. It was geometry that facilitated this step. The abstraction of eternal relationships was not supposed to apply to physical things, and no one, not even Galileo, was asserting that it did. He meant to apply geometry to the timeless (perhaps even divinely inspired) conception of Dante, but the step to the physical was so small that it happened innocently, by accident, almost without thinking. Dante's Inferno is, after all, by his genius made almost real. In this way the remarkable symmetries of Dante's mathematical universe and its uncanny realism acquired a new meaning. Dante commentary and mathematical physics had become, for a moment at least, the same thing.

11 Mathematics Old and New

A survey of Renaissance mathematics proper reveals a striking difference between its two subfields, the lively subject of algebra and the moribund subject of geometry. It is hard to avoid the conclusion that in the case of geometry, at least, philosophical preconceptions strongly limited what it could do. Algebra, on the other hand, is the success story of Renaissance mathematics. This branch of mathematics grew out of the arithmetic of the Arabs and Hindus, and it is not part of the classical legacy of Greece and Rome. It grew organically in a European context in response to real questions of commerce and finance. Algebra also gave rise to abstract problems like polynomial equations. Being outside the classical legacy, this new subject had no place assigned to it in Aristotelian philosophy. Polynomial equations had no philosophical overtones or meaning—they were just interesting and difficult problems. Perhaps for this reason, algebra in Europe developed rapidly and unencumbered from the fifteenth century on, and it looks like normal mathematics. There were discoveries of the kind that we still acknowledge today as real mathematics and disputes over priority, like the fierce and famous dispute between Niccolò Tartaglia and Girolamo Cardano over the solution to the cubic equation, the polynomial equation of degree three.

In fact neither Tartaglia nor Cardano was the first to solve the cubic equation. Piero della Francesca in the middle of the fifteenth century had tried and failed to find a solution, and Luca Pacioli, probably noting that Piero hadn't found it, declared that no solution could be found. Years later, faced with a deadline, Tartaglia inwardly cursed Pacioli for that careless and false remark. He had trusted Pacioli, he says, and had never seriously thought about this problem. Now, with little time left, he had every reason to believe that a solution existed after all, and that he had to find it or be disgraced.

The situation as far as we can reconstruct it was this. A professor of mathematics at Bologna, Scipione del Ferro, had found the solution to the cubic equation around the time of Piero's death, sometime in the 1490s, but he had kept it secret, for reasons that we will soon see. At the end of his life he had passed the secret on to his student Antonio Fior. Fior, in turn, had hoped to make his fortune with this knowledge by challenging the better-known Tartaglia to a mathematical competition, a kind of intellectual public spectacle and entertainment, the sort of event that could make or break reputations, perhaps ultimately garnering a professorship or losing it. Fior submitted to Tartaglia thirty problems, all of them cubic equations. Tartaglia in return submitted to Fior problems of various types, and the contestants then had a short time to prepare. It was now obvious to Tartaglia that there must be a way to solve cubic equations, but what was it? Without the solution he would be humiliated.

The possibility of a public challenge was the reason that a Renaissance mathematician might well choose to keep an important result secret. If Tartaglia had had something like this in his back pocket, he could have pulled it out now and given Fior thirty problems that *he* couldn't solve, but apparently Tartaglia didn't have any such secret ready to deploy. We have seen that Galileo kept his scaling theory secret for fifty years, perhaps for just this sort of emergency. That was undoubtedly why del Ferro had kept the solution secret too. Conceivably he had given it to Fior with instructions to publish it posthumously, but as we have seen, Fior did not do that.

The night before the public spectacle, Tartaglia found the solution. Humiliation had turned to triumph, as he was now able to solve all Fior's problems, and that is the last we hear of Fior. The formula for the solution to the cubic equation turns out to be surprisingly complicated. Students learn the simpler quadratic formula that solves quadratic equations, but they never learn the cubic one. It is just too complex. Tartaglia's audience didn't learn it either, because he continued to keep it secret, although he did promise to publish it someday.

This state of affairs was just too much for Girolamo Cardano, one of the most curious and prolific characters of the Renaissance. A physician and astrologer by training, he had discovered the pleasure of publishing books, in those early days of printing, and he published voluminously on medicine, astrology, mathematics, prophecy, gambling, and ghosts, among many other topics. He gained an international reputation, and in the course of his long life was even summoned to Scotland for a medical consultation. He ran afoul of a pope for casting the horoscope of Jesus Christ. And he was dying to know the solution to the cubic equation.

He pestered Tartaglia about it, and finally, for reasons that are not so clear, Tartaglia gave it to him. In return, Cardano had to promise that he would not publish it, as Tartaglia still intended to do so himself. In light of Cardano's propensity to publish everything he came across, one might expect that trouble was already on the horizon, but Cardano undoubtedly intended to abide by the agreement. Cardano had living with him a student named Ludovico Ferrari who was a gifted mathematician, far better than Cardano himself. As the two of them studied the new solution, Ferrari made new discoveries that built upon this one. In particular, he solved the quartic equation, the polynomial equation of degree four! This was a stunning result, never to be exceeded in the sense that the next such problem, the quintic equation, was proved in the nineteenth century not to have a formula for its solution.

Cardano and Ferrari were in a bind. They wanted to publish the new results, but they couldn't do that without publishing the old result, and this they had sworn not to do. As Cardano tells the story in his book, *Ars Magna*, the Great Art, where all these results are found, he and Ferrari in desperation had gone to Bologna and had managed to get access to the papers of Scipione del Ferro, and there they found the original solution. This, in Cardano's mind, freed them to go ahead, because they were not publishing Tartaglia's solution but del Ferro's. Cardano tells the whole story in *Ars Magna* with his customary garrulousness, including how Tartaglia had given him the solution, but Tartaglia's fears were realized. It is known in mathematics today as Cardano's formula.

Tartaglia was livid, and challenged Cardano to a public competition. Cardano was able to avoid this, perhaps because his position in society put him on a higher social level than the lowborn Tartaglia.[1] He sent Ferrari to meet Tartaglia instead, and the young man seems to have done very well. Tartaglia never got over this experience and peppers his last work, a kind of mathematical encyclopedia, with reference after reference to the putative mistakes of Cardano "and his creature Ferrari."[2]

Algebra, that is to say, was alive and well, even flourishing during the Renaissance, but Galileo never showed any interest in it. When he talked about mathematics, he always meant the classical mathematics of the Greeks, a completely different subject. In this respect, Galileo appears surprisingly conservative, ignoring what historians would consider the most exciting mathematics of his time. Galileo's education in the humanities is undoubtedly the reason for this preference. The humanities had nothing at all to say about algebra but assigned geometry the highest importance.

Mainstream Geometry

The classical humanistic legacy of science and mathematics in the Renaissance—astronomy, geometry, and so forth—was considered in philosophy to be almost a closed subject. Its problems were declared to be already solved, for the most part. There was, however, one little problem still left open, at least in its details, namely the calendar. Dante knew in 1300 that the Julian calendar isn't quite synchronized with the solar year, so that January will eventually be "un-wintered" (it will become a spring month).[3] Getting the calendar right was particularly important in setting the date of Easter correctly, and therefore the Church sponsored astronomical and mathematical studies throughout most of the fifteenth and sixteenth centuries. The Copernican idea is an indirect result of this work, as Copernicus participated in the project in his student days and continued his astronomical studies when he returned to Poland. This long effort culminated in the Gregorian calendar reform, implemented in 1582 (but much later, and at different times, in Protestant countries). In the collected works of Christopher Clavius, who oversaw the reform, you can look up the date of Easter next year, or a thousand years from now.[4]

As part of this long project Georg Peuerbach, beginning in the 1450s and then joined by his student Johannes Müller of Königsberg, known as Regiomontanus, compiled a table of the trigonometric sine, 5,400 entries in all, each one of them accurate to one part in 60 million.[5] Even twentieth-century scientists and engineers didn't use such accurate sine tables until tables themselves became obsolete with the introduction of electronic calculators. A spot check suggests that the computations are completely correct. It is truly impressive. This fifteenth-century computational feat is a significant achievement of mainstream Renaissance geometry.

It sounds a bit churlish to say it, after recognizing how painstakingly the job was done, but it really doesn't make sense to compute tables to this accuracy. This book-length trigonometric table contains, for practical purposes, no more information than a similar very small table in Ptolemy's *Almagest*. The only difference is that the new table works to many more places of accuracy—so many more that it took years to compile it. But measuring instruments of the time couldn't make measurements with anything close to this accuracy, so in a practical sense nothing was gained by all this computation. Apart from their use in applied problems, where the accuracy is limited by the measurements, these numbers have no significance. There is no pure mathematical reason to find them, unlike, say, the numbers found by Piero della Francesca in the *Libellus,* in connection with the Platonic solids, at about the same time. Piero did discover some things in doing his computations, as we have noted, but no one discovers anything in computing sine tables.

The systematic study of triangles was always a part of astronomy, and Regiomontanus, in his *de Triangulis* accompanying the sine table, abstracts this study, making it in effect a new mathematical subject, trigonometry. The work is not particularly original, however. It clearly derives from Arabic sources, even if the precise sources have not been identified.[6] The Arabs were, in fact, much more accomplished in this area, and an early fifteenth-century sine table computed in Samarkand, Ulugh Beg's *Sultani Zij,* went far beyond the European tables in resolution and accuracy.[7] Arab astronomy and trigonometry had benefited not only from classical Greek sources but also from Indian computational methods, including in effect the method that Newton would later deploy so effectively, infinite series. The Arab example forces one to ask, though, whether computational facility and painstaking improvement on classical results really was going anywhere.

The sine table is something that Ptolemy already knew how to compute, and in essence had computed in antiquity.[8] What Ulugh Beg and Regiomontanus were doing was just a slight variation on what had been done before, though to higher accuracy, a project in which the main difficulty must have been tedium. By contrast Copernicus's new plan of the heavens, *De revolutionibus orbium coelestium,* published in 1543, might have marked the beginning of an era of innovative thinking. In retrospect we would certainly say that it did. But at the time those few who actually used it saw Copernicus's plan as mainstream, just a variation on Ptolemy, another way to calculate. Copernicus's disciple Georg Rheticus, whose encouragement was partially responsible for Copernicus's decision to publish, spent much of his life on an even more elaborate and more accurate version of Regiomontanus's trigonometric tables. These appeared finally in 1596, when Galileo was already professor at Padua.

The geometrical side of Renaissance mathematics was still very much under the spell of Ptolemy some 1,400 years after *Almagest,* and not just in its construction of trigonometric tables. Regiomontanus's *Epitome,* a guide to *Almagest,* made Ptolemy's difficult book accessible and assured its continuing influence.[9]

Daniel Santbech

In 1561, just three years before Galileo was born, Daniel Santbech reprinted *de Triangulis* and the sine tables along with a book of astronomical and geometrical applications for them.[10] The book, despite its treatment of up-to-date topics like cannon shots, is much indebted to Ptolemy, beginning with a long philosophical introduction that repeats Ptolemy's tripartite understanding of the theoretical: theology, mathematics, and physics.

Santbech's book couldn't just dismiss practical questions as Ptolemy's did, though, because his book was necessarily practical. The sine table had no use apart from measured data, and after the philosophical introduction Santbech's first topic is how to build a quadrant, essentially a big protractor, for sighting and measuring angles. Connecting this artifact to the loftiness of philosophy posed a problem for Santbech that was only partially solved by his noting that Ptolemy had also described a measuring instrument. He gives his own hybrid subject of sighting and angle-measuring an impressive Greek name, *meteoroscopy*, literally "looking at high things."[11]

Like all mathematical books in the Ptolemaic tradition, Santbech spends considerable time on the philosophical importance of mathematics, now with a biblical twist. God was unwilling to see his creation ruined when he expelled Adam and Eve from the Garden, and so we retain our ability to discern the beauties of the world in the motions of the Sun, Moon, stars, and planets using mathematics (an ability that our first parents had to a much higher degree). The Devil, that deceiver, would dearly love to blind us to such beauty, of course, and distract us from it. In Santbech's sacred history of the world, the Devil's gradual success is evident in comparing the high place of mathematics in the past to its neglect today, but there is still hope, even in this, the senility of the world, that at least a few souls should long to understand the high truths. It is to such students that the book is addressed. (Oddly, in Santbech's sacred history of the Fall, the degeneration of the world, and its possible recovery by way of mathematics, the intervention of Jesus Christ is not even mentioned.)

Santbech's suggestions for things you could see, measure, and compute for yourself included practical astronomical examples, like telling time by the stars, but also deeply philosophical matters like the discrepancies in the calendar that had accumulated since Ptolemy's time. He was writing after Copernicus but before the Gregorian calendar reform, so these discrepancies were very lively questions. He refers quite impartially to Ptolemy and Copernicus and does not at all consider them to be in conflict. Copernicus had tried to incorporate some modern observations into his system, and that was a recommendation in his favor, but for many purposes his system was just an alternative digest of the same data that were already summarized in Ptolemy's scheme, and Ptolemy's theory was readily available in Regiomontanus's *Epitome*, so for theoretical questions Santbech refers to Regiomontanus. Tycho Brahe, as a young student at the protestant University of Wittenberg, would certainly have seen Santbech's book. It would have contributed to his conviction that astronomy was not on a sound foundation and that it needed a systematic modern data set, and it would even have helped him think how best to make new observations.

The earthly sections of Santbech's book tell you such things as how to determine the distances and heights of towers and, what is truly eye-catching in the charming woodcut illustrations, how to drop cannon-balls onto them. Santbech anticipates a lot of criticism from people who will say that he is "mixing heaven and earth" in treating this topic in a mathematical book, but he aggressively insists that he will do it anyway, without even making much of a philosophical argument about it.[12] "By Hercules, it is hard for me to close my mouth to this kind of person and to tolerate the pestilential license of such a deplorable . . ."[13] So begins his *ad hominem* counterattack on his still-hypothetical opponent, a fairly mild sample of his invective once he gets started. His defensiveness about using novel mathematics in an earthly setting is peculiar and edgy.

The main virtue of Santbech's new theory of cannon shots, from his point of view, is apparently that it uses trigonometry. It looks dubious, but Santbech insists that he has brought it to light by use of Euclid's *Elements*, "and I frankly confess I don't know of anyone who wrote it down before us."[14] Here is the theory: you should draw a straight line extending the direction of the cannon barrel as far as it takes for the cannonball to lose its "impetus," from which point the cannonball then drops straight down. Its trajectory is thus two legs of a triangle, a hy-

DE ARTIFICIO EIACVLANDI

A cannonball is imagined to travel by violent motion (straight) as long as it has "impetus," and to fall by natural motion thereafter. This figure makes a slight concession to the cannonball's falling even while it still has impetus in the alternative zigzag trajectory below the straight one. Daniel Santbech, *Problematum Astronomicorum*. Courtesy of Mount Holyoke College Archives and Special Collections.

potenuse going up, followed by a vertical drop going straight down. The goal of the cannoneer is to get the cannonball to drop onto the tower. Santbech knows that cannonballs follow curved trajectories, but he still argues that although cannonballs don't actually follow the straight hypotenuse of his constructed triangle, falling rather a bit below the hypotenuse, still, they stay very close to it, and when they get to the vertical leg, they then drop straight down along it, so that the geometrical theory is correct for the practical purpose of dropping the ball onto the tower.

Santbech emphasizes how precisely he can calculate the height and location of the vertical leg of the triangle (that is where the sine table comes in) even while he also points out that cannon are not precise. The memorable pictures that accompany his theory show only the geometrical construction, the cannonball going diagonally up in a straight line, then dropping down in a straight line,[15] even while the text concedes, a little reluctantly, that this is not quite what they do. The discrepancy does not bother Santbech, and he seems generally satisfied with the theory.

We might suppose that Santbech was just not much of a mathematician, and hence pay no attention to his simplistic and clearly inadequate theory, but a nearly identical theory of cannon shots was pub-

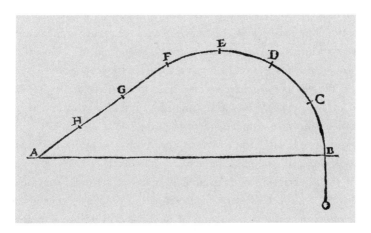

Tartaglia's version of the cannonball's trajectory is initially straight, then the arc of a circle, then straight down. Niccolò Tartaglia, *Nova Scientia.* Courtesy of Mount Holyoke College Archives and Special Collections.

lished at about the same time by none other than Niccolò Tartaglia in a little book called *Nova Scientia*.[16] We have seen from his other work that Tartaglia was a good mathematician, but when he deals with this subject, he somehow seems to lose his mathematical ability. Like Santbech he is satisfied with a level of proof that is almost nonexistent, even while his level of enthusiasm is curiously high. Tartaglia's work on cannon shots serves as a litmus to show us that there is something very strange about Renaissance geometry. Here we have good mathematicians ostensibly using geometry, but in a way that makes no mathematical sense. And this is almost within Galileo's own lifetime. Tartaglia was the teacher of Ostilio Ricci, the teacher of Galileo.

Even at this late date, professional mathematicians apparently thought of mathematics as an appendage of philosophy. As geometry, Santbech's theory of cannon shots is nothing. But as philosophy it makes sense. According to Aristotelian theories of motion, the cause of the violent motion of a cannonball would wear off over time, and left to itself the cannonball would assume its natural motion, which is falling straight down. Santbech's theory is not so much a mathematical theory of motion as it is a *translation* of philosophy into the language of mathematics. The two legs of the triangle, hypotenuse and vertical, are just the two kinds of Aristotelian motion, violent motion along the cannon barrel and natural motion toward the Earth, represented visually, geometrically. It must have seemed quite possible that the proper function of mathematics was as a language for philosophy, and that mathematics had nothing independent to add to what philosophy had already said. The reason for including the cannon example in Santbech's book is that the new sine table makes it possible to compute one aspect of this description more accurately than ever before. From this point of view, Santbech's theory makes sense, but only if mathematics has none of the power that we assume of it, and is only a passive means to express knowledge derived elsewhere.

New Translations

Although much of classical mathematics had already been translated into Latin by Dante's time in the thirteenth century from both Arabic and Greek originals, this work was repeated and extended in the fifteenth century. One incentive for making new translations was the arrival in Italy of large numbers of Greek manuscripts. The Council of Florence in 1439 had brought together representatives at the highest level of the Eastern Orthodox Church and the Roman Catholic Church, together with a large scholarly entourage, to discuss the ultimately unsuccessful reunification of the two major branches of Christianity. It was natural for scholars from the East to make use of the contacts they had formed at the time of the Council and to relocate, together with their books, when Constantinople fell to the Turks and the Eastern Empire collapsed just a few years later.

Another incentive for new editions was the invention of printing at about the same time. This might have meant just the reissuing of old translations, like Pacioli's 1509 Euclid, which was nothing more than Campanus's thirteenth-century translation along with Pacioli's peculiar attempts to explain what it meant. Tartaglia produced the first printed Archimedes in 1543, the one that Galileo studied so intensively on the advice of Ostilio Ricci, but it was just the thirteenth-century translation of Wilhelm Moerbeke. The widespread dissemination of such editions could only make it obvious, though, that the old translations were actually defective in some important places. There were substantive reasons to go back to original sources, insofar as that was possible, and prepare more faithful editions. Humanist bibliophiles knew how to handle variant manuscript sources, and their skills began to be applied to mathematical books as well.

Regiomontanus's *Epitome*, written in the 1460s but not printed until

1496, is in this tradition. It made *Almagest* accessible, but it also pointed out problems with the existing translation. For many purposes *Epitome*, as a cleaned-up paraphrase, replaced *Almagest*.[17] Regiomontanus formulated an ambitious program for printing new editions of all the important books of classical mathematics. It would have been a life's work even if he had lived to old age, but he died at the age of forty in 1476, the project hardly begun.

Over time, however, others brought out these books individually. The most important and influential of the humanist mathematicians was Federico Commandino of Urbino, whose Archimedes (1558) and Euclid (1572) were definitive. In particular his *Elements* V corrected Definition 5, the crucial definition of equality of proportion that was so hopelessly garbled by Campanus (and Pacioli). Commandino brought out printed editions of many other classical mathematical works that had been essentially unavailable, including Apollonius.

Commandino's best student was the young Guidobaldo del Monte, who on Commandino's death in 1575 completed his teacher's final editing project.[18] Thus when Galileo caught the attention of Guidobaldo, he was making contact with the very center of the humanist project in mathematics. Evidence recently pointed out suggests that this connection was extraordinarily important to Galileo, going even beyond what Guidobaldo was able to do in furthering Galileo's career.[19]

Galileo and Guidobaldo

Galileo had won the professorship at Pisa with Guidobaldo's support, and his frequent letters from Pisa kept up the scientific collaboration that they had begun when Galileo sent him his first theorems. They maintained a lively correspondence, and more than that, developed almost a father-son relationship, even though they probably had not met face to face.[20] At some point Galileo showed Guidobaldo his response to the Vitruvius story about Archimedes's eureka moment in the bath, something he had written as a little booklet called *La Bilancetta,*[21] although it was never published in his lifetime. He wrote it around 1586, and the material shows up as an entry in Guidobaldo's notebook for 1592.[22]

Galileo's view was that the Vitruvius story had to be wrong. Archimedes was not so stupid as to be surprised by the sight of water running over the side of a tub. Archimedes did, on the other hand, discuss the buoyant force on immersed objects and the conditions of balance. Galileo set himself to imagine how the real Archimedes, the one that he had come to know through his mathematical works, would have tested the crown. He proposes an elegant little device that combines two of Archimedes's results, the law of the lever and Archimedes's principle, which states that the buoyant force on an immersed object is the weight of the fluid displaced. In Galileo's device, the crown is balanced by a counterweight on a balance scale. Without being taken from the scale, the crown is now immersed in water, as a vessel is brought up from underneath it. Immersed in water the crown seems to lose weight, according to Archimedes's principle. Thus to bring the system back into balance, the counterweight must be moved inward on the opposite arm. The distance you have to move the counterweight, relative to the length of the arm, measures the density of the crown relative to the density of water. With this method you can measure the density of any convenient sample of anything denser than water, so you can make comparisons of the kind that solve the problem.

With a little care, such measurements can be made quite accurately, and Galileo actually did this. Throughout his career he refers to the known densities of things with great confidence, no doubt because he had measured them himself. His table of measured densities survives, and includes gold, silver, diamond, emerald, sapphire, and pearl.[23] Since the jewels cannot have been particularly large ones, he must have been clever in carrying out the procedure. Everyone who reads his little booklet is impressed not only with the theoretical idea but also with his suggestions for making the measurement, suggestions that take up much of the text. For instance, you must move a counterweight along one arm of the balance and accurately determine its new position. Galileo says that to do this you should first have wound fine steel or brass wire around the arm, very smoothly and evenly, so that when you need to determine the position you can count turns of wire to find the distance moved. And to do that more easily, he says, you can use a stylus, and drag its pointed end over the turns, counting the little bumps that you feel as you do this. The implication is that you wouldn't have a ruler to measure with, and that you would have to make it yourself.

Guidobaldo's notebook entries for late 1592 contain not only the essentials of *La Bilancetta* but other Galilean results as well, including the musical results that Galileo and his father had found in repeating the experiments of Pythagoras.[24] The appearance of several Galilean topics together was perhaps not the result of a spate of letters. It appears that Galileo was actually there. When Galileo moved from Pisa to Padua in 1592, Guidobaldo invited Galileo to make the journey via Montebaroccio,[25] and although there is no direct evidence of such a visit, the indirect evidence is very strong that Galileo not only stayed with Guidobaldo, but that what he saw there and what they did together marked a turning point in Galileo's career.

In *La Bilancetta* Galileo had described a measuring device of his own invention, but at Montebaroccio he would have seen an entire workshop in support of Guidobaldo's inventions.[26] The two of them shared an interest in measurement, and Galileo must have been very impressed with what Guidobaldo was doing. Once he had gotten set up in Padua, he put considerable effort into his geometric and military compass, as we have seen, installing a machinist in his own home for the purpose of producing it. Since he had not done anything like this at Pisa, it is plausible that he was inspired by Guidobaldo.

Guidobaldo was planning a book on perspective, and he and Galileo would surely have discussed it. The visit may even have inspired Guidobaldo to get to work on it, as most of the notebook after the entries relating to Galileo is devoted to perspective.[27] A letter of January 1593 from Guidobaldo to Galileo congratulates him on the terms of his Paduan appointment (which were apparently not final at the time of the visit) and describes Guidobaldo's excitement about his progress on the perspective book, for which he hopes he can get Galileo's criticisms.[28]

Most intriguing of all, it is possible that Guidobaldo and Galileo together performed an experiment that can plausibly be said to have led to the parabola law. This experiment is in Guidobaldo's notebook together with the other entries mentioned above.[29] After Galileo's arrival in Venice it shows up in the notebook of Paolo Sarpi, who quickly became one of Galileo's close associates.[30] Galileo had been working hard to understand motion during his years at Pisa (whether he dropped balls from the Leaning Tower or not), writing Latin drafts of treatises that owed much to Aristotelian and medieval notions. Guidobaldo was not sympathetic to such approaches, and in disputing about it they may have decided to make an experimental test. The experiment they describe is rolling a ball across an inclined plane, perhaps a slanting roof. The path of a ball on such a plane is, apart from the small effects of roughness and friction, a parabola, although they wouldn't have known that. They might have begun to suspect it, though. The most peculiar thing, in their view, was that the path was symmetrical going up and coming down. That is very different from the theoretical philosophy that says, in the versions of Santbech and Tartaglia, that the path should be a hypotenuse going up and a vertical coming down.

Galileo's experiments with balls rolling down inclines at Padua are one-dimensional versions of this two-dimensional experiment, under better control and with fewer experimental variables. Eventually he would synthesize the Paduan experiments with a theoretical idea on horizontal motion to arrive at the parabola law, but perhaps this synthesis was already present in his mind from the beginning, because he had seen the result at Montebaroccio with Guidobaldo before he ever got to Padua.

12 Transforming Mathematics

Galileo's background in the humanities and the arts was formative for his approaches in mathematics and the emerging sciences. That is the idea I began with, and the idea I return to in this chapter. In particular I consider the importance of humanist history and literature for his understanding of the mathematical legacy, and the importance of standard artistic practice for what he was able to do in mathematical science. Neither history, literature, nor artistic practice were elements of a normal mathematical education, so any benefit he obtained from such a background was virtually unique to him. Conversely, a normal mathematical education included philosophical stances that Galileo never adopted.

Humanist Insights

As part of the general humanist project of recovering the classical past, nonmathematicians, bibliophiles, and classicists worked with mathematical texts just as they worked with literary and artistic texts, noting cross-references, establishing chronologies, and creating a history. Apart from better and better translations, this work of cultural integration did not add anything to the ancient scientific materials in the mathematical or scientific sense, but it did reveal how they were related to each other, and it hinted at how to make sense of them, and ideally how to pay attention to the parts that mattered most.

It was ultimately important to realize, for example, that the Romans had not done any mathematics at all, and that under Roman rule even Greek mathematics had degenerated. This observation, to anyone who understood it, would have been enough to focus attention on Hellenistic mathematics before Roman rule, rather than (for example) on computing sine tables that were essentially already available in Ptolemy. Just to understand this much, however, required an expanded understanding of what it meant to "do" mathematics, and that was just the concept that was elusive. The best humanist scholarship still could not, by itself, distinguish good mathematics from mediocre mathematics, or even say what mathematics was. The legacy was curiously silent on that crucial point.

When we say classical, we usually mean Greek and Roman together, as if they were two complementary halves of a whole, something that the Romans also liked to believe.[1] The simplest and most natural humanist position was not to make invidious distinctions in the legacy or to favor one part of it over another. An interesting exception is Vincenzo Galilei's scholarship in music, which makes a clear distinction between the Greek and Roman contributions, and describes them quite asymmetrically. Galilei more or less says that the Romans killed Greek music:

> The Greeks, masters and inventors of music—along with all the other sciences—rightfully held it in great esteem. Since it is a delight of life and also useful to virtue, the best legislators ordered that it should be taught to those who were born to attain perfection and human blessedness, which is the goal of a city.

But in the course of time, the Greeks lost music and also the other disciplines along with their empire. The Romans had a knowledge of it, taken from the Greeks, but they exercised principally the part suitable to theaters in which tragedies and comedies were performed without appreciating much the speculative part. Continually occupied in war, they did not pay much attention to this side of music and so easily forgot it.[2]

The quotation is from the very beginning of Galilei's *Dialogue*, but it must reflect his conclusion, his impatience with his contemporary Zarlino and the Roman-era authority Ptolemy, and his rediscovery of the theory of Aristoxenus, so much less constrained, an early Hellenistic point of view better adapted to describe the richness of musical possibility.

An awareness of history in the Galilei household may have helped young Galileo to notice, just a few years later, that a Roman retelling of a Greek story, namely Vitruvius's story of Archimedes in the bath, was defective. Consistent with his father's disillusioned view of the Romans, Galileo believed that the Roman Vitruvius hadn't really understood Archimedes. The method described by Vitruvius was probably not even sensitive enough to make the measurement, he suggested. Galileo's *bilancetta* was a far more likely solution to the problem of the crown. Galileo must have been surprised (but perhaps not completely surprised) to realize, in view of the high authority ascribed to all classical authors, that Vitruvius knew less about Archimedes than he did. An experience like this could only have confirmed his interest in the earlier Greek materials over the later Roman period ones.

The Pythagorean experiments on tension and musical pitch that Galileo conducted with his father were perhaps the most striking example of all. The Roman period sources were wrong, but the Greek Pythagoras had been right: the weights do control the pitch by a kind of proportionality. The two of them rediscovered by experiment what the early Pythagoreans must have done. Here, in a quantitative experiment, guided by a mathematical theory, was a method and a source of knowledge that could be generalized, perhaps what Galileo meant by Pythagoras's "method of philosophizing."

It is worth noticing that Pythagoras and Archimedes, Galileo's two favorite philosophers according to Gherardini, are exactly the two that are implicated in his first scientific efforts, and that each of these efforts involved both mathematics and experiment. It is also worth noticing that both these problems originated in classical stories, not the first source we might think of for scientific questions, but Galileo's first sources nonetheless.

Another early problem from literature had almost proved a disaster, Galileo's use of Dante in his Inferno Lectures. At some point he recovered from this near debacle with the scaling theory finally published in *Two New Sciences* that used simple geometrical insights on the behavior of areas and volumes when they are scaled up. This too might well have qualified in his mind as "Pythagoras's method of philosophizing" and must have confirmed his confidence in such an approach at an early age.

Evading Ptolemy

As significant as what Galileo learned in the humanities and the arts is what he did not learn: namely Ptolemy's dictum that mathematics applied only to the heavens. In the context of what Galileo knew about mathematics this notion made no sense. Young Galileo knew the perspective construction, for example, and later consulted with Guidobaldo del Monte on his perspective treatise. Contrary to the opinion of Aristotle, insisted upon by Ptolemy, perspective is a mathematical theory that does describe things on Earth after all. There could be no doubt about that. Whatever Aristotle had said, the transitory nature of things on Earth does nothing to spoil the theory of perspective. Perspective theory had never fit into philosophical mathematics, in fact. In the late fifteenth century it was loudly asked whether the quadrivium shouldn't be expanded to include perspective as a fifth mathematical subject. It couldn't be explained away. One has only to look at a perspective painting to confirm it by personal experiment.

Another corrective to Ptolemy's dictum is the phenomenon of Archimedes. In one view, Archimedes's mathematics and his spectacular engineering feats have nothing to do with each other. Plutarch, for example, passes along the Roman view that although Archimedes had built unsurpassed war machines, he scorned to say anything useful, leaving behind only mathematics. Yet the surviving lore about Archimedes, taken by itself, as if Ptolemy (and Plutarch) had never said anything about it, suggests no divide at all between mathematics and its applications, but rather just the reverse.[3] And this was basically how Galileo did come to Archimedes, not through the Ptolemaic tradition of mathematics and astronomy, but through the artistic tradition. From this point of view, it must have seemed natural to try to reconstruct one of Archimedes's machines, and to relate it to the abstract treatises, as Galileo did in *La Bilancetta.*

By way of comparison it is instructive to see how other mathematicians regarded Archimedes. It is remarkable how many different ways there were to interpret him. He had been translated in the twelfth and thirteenth centuries, along with Euclid and Ptolemy, but by the age of printing, interest in him seems to have lagged somewhat. Piero consulted and made use of the Moerbeke translation in manuscript, but the first printed Archimedes did not appear until 1543, decades after the first printed Euclid in 1482, or Regiomontanus's *Epitome* of Ptolemy in 1496. That Archimedes does not deal with astronomy may have made him difficult to categorize. Ptolemy's *Almagest* makes use of Euclid, but not of Archimedes, whose interests are more explicitly down to earth.

Daniel Santbech, the Protestant author of a 1561 book on geometry and astronomy, had to justify the sections of his book that deal with earthly geometrical problems to an audience that was not used to seeing astronomical and earthly topics treated together. He does this by invoking stories of Archimedes, even though he doesn't use any Archimedean mathematics. He mentions the siege of Syracuse, including the burning mirrors, the planetarium, the story of how Archimedes and one of his mechanisms moved a ship that the whole citizenry of Agrigentum could not budge, and the Archimedes screw for lifting water.[4] Just before treating cannon shots, the most controversial of his earthly sections, he brings in one last Archimedes story for a more extended treatment, the story of the goldsmith and the crown. That story has absolutely nothing to do with cannon, of course, but Santbech is letting us know what Archimedes means to him: not mathematics, but rather license to consider earthly problems.

Galileo's contemporary Kepler was quite uninterested in Archimedes in his early work. The only book he seems to know is *The Sand Reckoner*, which is also the only surviving work of Archimedes that touches on astronomy. In his 1604 *Optics* he discusses the balance, but without even mentioning Archimedes, something that seems almost impossible. Archimedes's treatise on the balance was pivotal for Galileo: Kepler ignores it. In *Astronomia Nova*, though, Kepler suddenly found a useful idea in Archimedes's *On the Measure of the Circle*, and it made possible his formulation of his Second Law, the one that refers to the area of the curved orbit. This conception of Archimedes was fresh in his mind when, a few years later, he contemplated the volume of wine casks in the *Nova Stereometria*, which begins with what Kepler considers an extension of the results of Archimedes on areas and volumes to new figures of revolution. It must be said that although Kepler cited Archimedean results in these works, he pointedly criticized Archimedean methods, comparing them to a thorny thicket, clearly implying, as Plutarch had done long before, that Archimedes's mathematics was too abstract and unintuitive. His view of Archimedes is restricted to the one thing he had found useful, the computation of areas and volumes, and his method is actually not at all that of Archimedes, as contemporary critics pointed out, but an intuitive method, much more flexible than that of Archimedes but ultimately relying on him.[5]

The issues that underlie the reception of Archimedes are, in a way, still with us. In the mid-twentieth century it was suggested by Alexandre Koyré that Galileo could not have performed real experiments, because his physics was Archimedean, by which he appears to mean purely theoretical, à la Plutarch:

> An Archimedean physics means a deductive and "abstract" mathematical physics, and it was just such a physics that Galileo was to develop at Padua. A physics of mathematical hypotheses; a physics in which the laws of motion, the law of fall of bodies, are deduced "abstractly," without involving the idea of force, without recourse to experiments with real bodies. The "experiments" which Galileo, and others after him, appealed to, even those which he did actually perform, were not and could never be any more than thought experiments.[6]

This provocative suggestion was taken seriously until Thomas Settle demonstrated that Galileo's accounts of his own experiments are quite consistent with modern recreations of them.[7] That is, they seem to have been real experiments after all. But it is startling to notice that Koyré articulates essentially the same belief in a Ptolemaic separation of theory and practice in Archimedes that Plutarch recorded over two thousand years ago, and in a book called *Galileo Studies*, of all places. Koyré, through this misunderstanding, inadvertently highlights the subtlety and insight of what Galileo did.

Experiment and Artistic Practice

Galileo, right from the start, was able to use measured data intelligently and meaningfully in spite of its imprecise and variable character. This variability always surprises students who are doing experiments for the first time, although they know in principle that this is what happens. It is not enough merely to have been told about it. There is no substitute for doing an experiment yourself when you don't know what the answer is supposed to be, relying on numbers that scatter about somewhat randomly. It is not enough to read about it. The actual experience is shocking. The most natural reaction is to give up, because the process seems to be too flawed to be reliable. Clearly the measurement process is tainted with error. Students often assume that the error is some kind of mistake that they are making and that it is their fault. They implicitly think that since we do have theoretical science and experimental science after all, then someone else must be able to do it correctly, make perfect measurements. Of course that is not the case. We have precise theories and imprecise data, and somehow we put them together. It is not easy to say how this works, but Galileo taught himself to do it, and he was one of the first to understand this very subtle process. This is how the theoretical and the practical confront each other.

The earliest example of Galileo's use of real data is a table of measurements that accompanies *La Bilancetta*.[8] It is in Galileo's own hand and must have been for his own use. Weights are recorded to $1/32$ of a grain. Even for his smallest sample, a sapphire weighing $5\frac{3}{4}$ grain, this would be better than a 1 percent measurement if it were really possible to distinguish differences as small as $1/32$ grain, but the instrument was actually not that accurate. Galileo records enough measurements that we can see that the measurement uncertainty was more like 2–3 percent. He doesn't show repeated measurements on the same sample, but he does include two samples each of several materials (gems). For example, a ruby weighing 169/16 grain in air weighed only 127/16 grain in water, while a second ruby weighing 491/10 grain in air weighed only 355/16 grain in water. If the rubies had had the same density, and if the measurements had been perfect, the two ratios of the weights in water to the weights in air would have been the same, but they are different by more than 4 percent.

The table is very revealing. First, for several materials Galileo used two samples, as if it were not enough to make a single determination, however careful. That the second measurement invariably did not quite confirm the first measurement, and that he nonetheless included both measurements in the table, shows an awareness and acceptance of the variability in the measurement that is worth noticing. One column in the table is used to scale up the measurements so that the materials can be compared, as if all the weights in air had been 576 grains. Galileo only scaled the first sample in each pair, perhaps taking that first one as the standard. Not scaling the second one, as if to avoid the embarrassment of seeing how different it was, is a bit odd, but psychologically not hard to understand: we tend to believe the first measurement, and to compare subsequent ones to it. Experimentalists learn to avoid this trap. Here we can see Galileo struggling with it. It is as if he both did and did not want to see the second measurement. Finally, and most important, he never for a moment doubted the usefulness of the notion of density, even though a naive reading of the table might interpret it as an experimental refutation of the notion of density. That is, he could accept the variability in the measurement, differing from what the theory of density predicted, and yet reconcile these things in his own mind. That is the trick, and it is not easy. It is what Koyré said was impossible.

The inherent error in measurement and how to deal with it is the first topic taken up on the Third Day of Galileo's *Dialogue Concerning the Two Chief World Systems*, probably the first extended discussion of this point in any scientific context. All three participants are eager to discuss a book that Simplicio has brought for them, the *Anti-Tycho* of Chiaramonti, which purports to prove with Tycho's own methods that Tycho was wrong in locating the new star of 1572 above the Moon, where according to Aristotelian philosophy no change should ever occur. Chiaramonti uses measured sightings reported by observers all across Europe to triangulate the position of the new star, and he does indeed find that it must have been closer than the Moon, not farther, and hence that Aristotelianism is safe. Salviati, however, who has been perusing the book overnight, is scathing in his criticism. Chiaramonti, he says, seems to be "expecting such gross ignorance on the part of everyone into whose hands his book might fall that it quite turns my stomach."[9] Out of all the observations, he has selectively chosen just those pairs of observations that put the new star close to the Earth, and has systematically ignored all the other pairs, much more numerous, that give enormously greater distances. Galileo argues that one must consider the entire data set, knowing that it is, in a sense, self-contradictory, and extract the most plausible meaning from it. This discussion documents Galileo's sophistication in dealing with the messiness of real data (and these data, taken by different observers of varying skill, are *very* messy). The naive point of view is put into the mouth of Simplicio. As Salviati questions him, they agree that if everyone had measured the distance correctly, then they would all have gotten the same value.

> *Salviati:* But if, of many computations, not even two came out
> in agreement, what would you think of that?
> *Simplicio:* I should judge that all were fallacious, either through
> some fault of the computer or some defect on the part of

the observers. At best I might say that a single one, and no more, might be correct; but I should not know which one to choose.[10]

This is a joke, but Galileo's readers must have realized that they were also laughing at themselves, because a more sophisticated view of statistical variability had never been articulated.

Although there was no theory of statistical variability in Galileo's day, there was still a history of perfect theory confronting imperfect practice in the mathematical theories of the arts, where the problem was handled in a purely pragmatic way. Whatever one's theory of tuning in music, for example, whatever music theorists might say, the act of tuning in real musical performance is limited by the capabilities of the human ear. Perfect tuning is not realistic or even possible. It only has to be good enough to satisfy the ear. As we have noticed, this point is made even in Plato's *Republic*, although in the service of the opposite idea: perfect proportions. Vincenzo Galilei was the most outspoken theoretical advocate for tuning by ear, and not by perfect arithmetic.

Piero's perspective treatise used Euclidean geometry, but Piero was explicit about the imperfection of what he was doing, compared to the perfection of the theory. He recalled Euclid's definition of a point as "that which has no part," but pointed out that for his purpose a point is the smallest ink dot that you can make, which is not the same as a geometrical point. A geometrical line has no breadth, but Piero points out explicitly that his lines do have breadth. Theory and practice did not correspond perfectly, but that was no reason to give up on the construction. No one ever argued that. The correspondence only had to be good enough to produce a satisfying visual effect.

Galileo's easy acceptance of the imperfection of experiment juxtaposed with the perfection of the corresponding mathematical theory must have been made much easier by his experience in the arts, where such accommodation was standard artistic practice.

Galileo's view of experiment is not that it tests theory but that it is the foundation of theory. Experiment suggests the principles that theory should use; I will say more about this in the last chapter. In *Two New Sciences*, Galileo talks about this casually, as if it were something that everyone knew and accepted, a rhetorical device that should be appreciated for the fiction that it is. On the Third Day of *Two New Sciences*, Simplicio, the Aristotelian philosopher, in delicious subservience to the Author, actually asks to see an experiment on accelerated motion:

> *Simplicio:* I am still doubtful; and it seems to me, not only for
> my own sake, but also for all those who think as I do,
> that this would be the proper moment to introduce one
> of those experiments—and there are many of them, I un-
> derstand—which illustrate in several ways the conclusions
> reached.
>
> *Salviati:* The request which you, as a man of science, make, is a
> very reasonable one; for this is the custom—and properly
> so—in those sciences where mathematical demonstrations
> are applied to natural phenomena, as is seen in the case
> of perspective, astronomy, mechanics, music, and others
> where the principles once established by well-chosen ex-
> periments, become the foundation for the entire super-
> structure.[11]

Salviati names as customary experimental sciences the two quadrivium sciences (music and astronomy), perspective, and mechanics. That was not the traditional view of these things, of course. Galileo's father had contributed to music as an experimental science, and Galileo himself had done a great deal to redefine experimental astronomy in his pioneering use of the telescope. These quadrivium sciences were traditionally regarded as almost purely mathematical, not experimental, or if they were experimental, it was only in the service of truths already known about them, in filling in the details. Galileo here is talking about new foundations for these subjects on the basis of experiments in which he had taken part, a revolution that not everyone had accepted by any means. A similar breakthrough had taken place in perspective in the fifteenth century, as Galileo implies, when it became a mathematical science founded on experiment, but the revolutionary implications for mathematical physics were not immediately appreciated at that time by anyone. As for mechanics, it is not what we now call mechanics, one of the "new sciences" of the title—that would be circular—but rather the technology of mechanisms, based on Archimedes's law of the lever. It is a little odd that Galileo cites mechanics as an example of experiment supplying the foundation of theory, because Archimedes's proof of the law of the lever seems to be entirely abstract, reasoned from axioms and not experiments. Galileo tacitly assumes what was undoubtedly obvious to him, that Archimedes's axioms were grounded in experience, and that experiment and theory in Archimedes are intertwined.

Observation and Artistic Practice

Galileo was an artist trained in both theory and practice, and he was also the first to observe many remarkable phenomena. These two circumstances may be more than merely coincidental. It seems plausible, and perhaps almost obvious, that someone who is trained to see, and who thinks about the process of seeing, sees more and sees better. Contemporaries, using Galileo's own telescope under his supervision, still, in numerous cases, could not see the moons of Jupiter. Unable even to see them, they could certainly never have discovered them. Perhaps the best documented example of Galileo's practiced and educated eye is his interpretation of the light and dark regions of the moon when he first observed it through a telescope in late 1609. Accustomed to thinking of how to model three-dimensional figures under various conditions of lighting, he recognized, in the patterns of light and dark, mountains and craters. Contemporary observers like Thomas Harriot were unable to do this.[12]

Galileo's knowledge of perspective theory was all that he needed for a quantitative interpretation of what he saw in the telescope. The apparent size of things, according to optical theory, is really their angular size. Thus if things look bigger, it must be that the telescope bends the visual rays to make them enter the eye at a larger angle. This is Galileo's entire theory of the telescope, as he describes it briefly in *Sidereus Nuncius*. How the telescope accomplishes this is something that he never investigated further, nor needed to. As for making improved telescopes, he must have done it by making small changes, noting their effects, and extrapolating to the best design he could manage. This process requires very little mathematics, but it does require careful observation, together with quantitative assessment of what he observed. These are the skills of a working artist.

An incident from late in Galileo's life, when he was already going blind, makes this point unmistakably. By that time thousands of people had observed the Moon attentively through telescopes, but it was still Galileo who first noticed that the Moon librates, that is, it rocks slightly back and forth, showing us a bit more than one hemisphere if we observe it carefully over time.

The craters on the Moon are named for the great scientists and mathematicians of the past. It seems a bit unjust that Galileo, who first recognized them for what they are, is honored with a very obscure crater near the western limb, one that is even hard to make out. It is some consolation to consider that this feature is very well placed to show us libration, as it moves now closer, now farther from the edge, to remind us of Galileo's last astronomical discovery.

The Parabola Law

The parabola law is simple enough, in a formal sense, that one can find what appear to be special cases of it hundreds of years before Galileo. The French physicist Pierre Duhem, around the turn of the twentieth century, suggested that it had been discovered already in the fourteenth century, 300 years before Galileo, and that Galileo and Copernicus were nothing but "the heirs and, as it were, disciples of Nicole Oresme and Jean Buridan."[13] He was pointing to a formal similarity between certain medieval arguments and Galileo's arguments. A corrective to the enthusiasms of Duhem is provided by Annaliese Maier, who points out how abstract the fourteenth-century theory was, and how far from making statements about any experiment.[14]

In assessing this situation, it is necessary to remember that formalism is one thing and meaning is another. It was the meaning of mathematics that changed over the Renaissance, not the formalism. The medieval theory was not so much a theory of motion as it was a theory of *any* quality that varied with position or time, the quality of color, for example. That it happened to be useful in a later theory of motion is perhaps just a striking example of the flexibility of abstraction.

What distinguishes Galileo's parabola law from the work of his predecessors, and from mere formalism, is that it reveals a true understanding of the phenomenon of fall, not just the formal consequences of assumptions in a logical argument modeled on geometry. An important reason for this is that he merged mathematical reasoning with experiment and measurement, in the ways just sketched. Galileo gets every relevant thing right, not necessarily by way of a logical argument but in a physical sense. This took a long time, because the words "relevant" and "in a physical sense" have a meaning that can only slowly be appreciated and understood.

Galileo had lived with the problem of accelerated motion for fifty years. He had considered it from every point of view that he could think of, done ingenious experiments, argued for erroneous positions, and only years later noticed and corrected inconsistencies. None of these things by itself was decisive. As we have seen, Galileo may have had reason to believe in the parabola law as early as 1592, when he visited Guidobaldo del Monte.[15] But to build a framework in which this law made consistent sense with everything else that he knew—that was the work of a lifetime. In the end he knew it was right, in some important sense, because he had gone over it so often, so critically, in so many contexts, and it all held together.

In this respect, Galileo's work had the character of modern science, in not ever being strictly proved but in being good enough. This subtle understanding of what mathematical physical law should look like, quite different from pure geometry and even from the modern popular conception of science, is perhaps Galileo's subtlest contribution. Working scientists know that both theorists and experimentalists in fact go over and over the same ground, repeating what they have already done countless times, occasionally noticing here and there something they had not noticed before, making improvements in technique, streamlining the process. Gradually things become more consistent and clear. Some difficulties disappear without ever being explained or understood. Gradually one gains confidence in a coherent body of theory and practice that works consistently. This is what Galileo had done with the parabola law.

Even after publication, Galileo wasn't satisfied with his treatment of the parabola law. He was confident of it, but it still wasn't finished. He continued to tinker with the argument, as he describes in a letter of December 3, 1639, to his former student Castelli. Young Viviani, who had just come to live with him, had studied the proofs, and some of his questions and objections had stimulated Galileo to find a new argument, which Viviani had to write out for him due to his blindness.[16]

GALILEO WORKED ON THE problem of motion for fifty years. As he might well have said, "When I get an idea (provided it is a noble and interesting one), if I realize that it is beyond my powers, I am all the more eagerly inflamed—I tingle to my fingertips. I'll leave no stone unturned to go after it—there is nothing I desire more ardently."[17] These words, which seem to suggest long experience with a difficult and worthy problem, were supposedly spoken by a much younger man at the beginning of his career, someone perhaps even too young to appreciate their real meaning, but they describe Galileo's pursuit of the parabola law perfectly. The intriguing possibility that they were actually written by Galileo after all, in a previously unsuspected mathematical manifesto, is the subject of the last chapter.

13 The Oration

In August 1627, Galileo wrote his last letter to Kepler after a silence of seventeen years. The letter itself does not amount to much, just a recommendation for a young man named Johannes Stephan Boss, otherwise unknown, but in a postscript Galileo says, "I send, enclosed with this letter, an oration of Niccolò Aggiunti, an outstanding young man in serious and humane letters: I am sure that you will read him with great pleasure, and that you will find him wonderfully to your palate and taste."[1] In view of the complicated relationship between Galileo and Kepler, it is natural to ask what this little book is, and what Galileo could have meant by sending it to Kepler.

Niccolò Aggiunti was professor of mathematics at Pisa, having taken the position the year before at the age of twenty-six. His *Oration in Praise of Mathematics* seems to have been in fulfillment of an inaugural duty in assuming the professorship, "to get his lectures off to a good start," as Marcantonio Pieralli, Moderator of the Ducal College, says in a short dedicatory preface.[2] If the little book had really been to Kepler's taste, it would have been about mathematics and astronomy, but this is where the strangeness begins. It hardly mentions astronomy. Did Galileo perhaps mean to broaden Kepler's horizons?

To praise mathematics in 1627 but to leave astronomy out is more than unconventional. It is unheard of. So what does the Orator mean by mathematics? Conceivably the oration could have been about merely practical mathematics, but the Orator disposes of this idea in a few lines right at the beginning:

> Who disparages mathematics? I reply, a great many people do: for empty and ridiculous praise is disparagement, and not praise, and someone who praises faintly and unintelligibly (if we believe Favorinus) is worse than someone who criticizes openly. Who is not angry at the inept and ignoble praise that certain pretentious people bestow on this excellent subject? I mean those who pass themselves off as geometers, considering it to be the essence of the mathematical art to measure the height of towers, the depth of ditches, the areas of fields, and the positions of places, and to confine so narrowly and ignobly this science that ranges over the breadths of the Earth and Sea and whose realm extends to the limits of the starry sky. This is as bad as saying that sculpture was invented to make fine pots and ladles, or that Pythagoras painstakingly worked out the musical modes so that a shepherd could entertain his flock, pasturing in the sun, on his many-holed boxwood flute.[3]

(The Orator's casual way of associating mathematics and the arts might already have gotten our attention.) What he means by mathematics is developed at more length in a wide-ranging essay of more than thirty closely printed pages, and it turns out to be none of the usual meanings of mathematics, but something much more radical: a philosophy of earthly things, grounded in geometry and arithmetic, what we would now call a mathematical physics, except that mathematical physics did not exist in 1627. The Orator's principal examples come from the arts, which he seems to know as if he were a master of them, and from the stories of Archimedes. This does not at all sound like a beginning Renaissance mathematics professor. On the contrary, it sounds exactly like Galileo himself.

The Oratio *as a Galilean Work*

In 1626 Niccolò Aggiunti had been a devoted follower of Galileo, on the most intimate terms with him for four years, time enough to have absorbed Galileo's thought and drawn upon it for his oration. Of all Galileo's students, Aggiunti seems the closest to him, the one who perhaps most readily absorbed and reflected his attitudes and ideas.[4] There are many places, though, and not just the one quoted above, that sound not merely like Galileo's thought, but like Galileo's thought in his own words. This would be impossible to prove, but to argue that it is at least possible, I can imagine the following scenario. Galileo, eager to publish a certain philosophical view of mathematics but not eager to be attacked (or counterattacked) for it, organizes and writes out, in Italian, most of the *Oratio*, which Aggiunti, an accomplished Latinist, then translates into Latin. Together they find the countless classical allusions that illustrate the encyclopedic content of the work. It is worth noting that after 1626 Aggiunti occasionally completed Latin translations for Galileo, including the Letter to Cardinal Orsini on the flux and reflux of the sea, and at least two letters to foreign scholars. This practice might have begun with the *Oratio*, which would therefore have a dual authorship, in spite of what the title page states.

Pieralli's preface says that because of the constraints of custom and time, the oration that was given was lacking many sections, and it was "despoiled of much illustrative material." That is, the oration that Aggiunti gave was much shorter and plainer than what was ultimately published. Someone put additional work into the printed version. Pieralli even tells us more about that. The little bit that Aggiunti delivered as his oration aroused in his audience a great desire to read the whole of it, but Aggiunti was too modest to let them see it. When Pieralli was in Rome some months later, however, at the home of Giovanni Ciampoli,[5] he was asked again and again about that oration of Aggiunti. Why did he keep the lovers of the fine arts in such suspense? Advised of these kind requests, and trusting that he would be protected by the patronage of those who made them, Aggiunti at last acceded to their wishes. That is what Pieralli says.

Remarkably, there is another account extant of that occasion at Ciampoli's house in Rome, and in this version the request was a different one.[6] A letter from Ciampoli to Galileo, dated July 10, 1627, describes Pieralli's coming to his house in Rome. Everyone there wanted news of Galileo, his state of health, and the progress he was making on his *Dialogue Concerning the Two Chief World Systems,* a work they were all eagerly anticipating. Mentioned by name in the group is Galileo's former student Benedetti, who had been Professor of Mathematics at Pisa before Aggiunti, and under whom Aggiunti had studied. Aggiunti and his oration were not their concern, however, contrary to Pieralli's account. Rather the assembled group was unanimous in making two requests: (1) that Galileo should right away let them enjoy some part of the work that he had finished, and (2) that the good wishes of his friends should persuade him to finish the *Dialogue.* That is what Ciampoli's letter says. Pieralli had said that the oration was published in response to the gathering at Ciampoli's, but that account only makes sense if the *Oratio* is Galileo's, not Aggiunti's, because the group wanted a work from Galileo. Pieralli's introduction would be a sly nod to those in the know, apprising them that Galileo was the real author.

It would be entirely in character for Galileo to write a public speech for one of his protégés to deliver as if it were his own. He had done exactly this just a few years before, assigning Mario Guiducci to give an address on the comets of 1618 to the Florentine Academy, an address that was actually written by Galileo and not Guiducci.[7] As Galileo says in his 1623 book *The Assayer,*[8] he is tired of being attacked for his discoveries. Guiducci's name was meant to deflect any attacks on the subject of the comets, apparently, and Galileo is plainly irritated that his enemies have found him out. Aggiunti's oration, on the other hand, although it is a vicious attack on the philosophy faculty, among other things, seems to have provoked no comment at all, perhaps because the cover story held, the venue was obscure, and young Aggiunti was not worth a public reply.

Aggiunti was a client of Galileo's in the sense of the patron-client relationship, and he addresses Galileo in letters as his patron, even if it is sometimes with affectionate informality. He would not have had the professorship if Galileo had not given it to him. Galileo, as the senior professor of mathematics at Pisa, among his other titles, had preferred Aggiunti over another of his students, one whom history has judged by far the better mathematician, Bonaventura Cavalieri. When Aggiunti made the oration, it cannot be that Galileo was uninvolved.

A letter from Aggiunti to Galileo at about the time of the oration shows concern about Galileo's lack of progress on the *Dialogue*, the same concern expressed by the group at Ciampoli's, and goes on to say,

> Other people come to the University to learn, but if I wanted to learn I would have to leave the University and come back to you. Since I have been here I have learned nothing, nothing *at all*; and from this I draw two conclusions. One is that I know a lot, because there is no one here who can teach me. The other is that I am ignorant and worthless, because although there are so many millions of things to discover, I don't discover even one. And this second one is the truth, and because of it I live in continual torment.[9]

Aggiunti closes this letter by declaring, "I can't think of anything else to say, except that I would gladly occupy myself with any command of yours, because anything undertaken to please you is a relief and consolation from my other duties." We can be sure, for example, that if Galileo had asked Aggiunti to prepare his oration for publication, Aggiunti would have done it. But according to Pieralli, only the request from Rome made him do it. This account, in which Galileo is never mentioned, is suspect. Galileo was delighted with the final version, since he sent it to Kepler, so it seems unlikely that he had left it to other people to push for its publication. Aggiunti would have had no need of either requests or protection from Rome. The far more likely explanation is that Galileo was involved but concealed his own role. The very short time, just a few weeks, between the gathering at Ciampoli's and the printing of the *Oratio* suggests that Pieralli had brought the manuscript to Rome with him, ready for publication.

The *Oratio* is dedicated to the Grand Duke Ferdinand II, but not by Aggiunti. Rather Pieralli writes the short dedication, as a representative of the University of Pisa, in gratitude for the generosity of the Medici in supporting the university, offering to the grand duke this token of the fruits of scholarly endeavor. Aggiunti, from the time of his doctorate in 1622, had been a courtier at the Medici court, where he probably tutored the boy Ferdinand II (it is known that he tutored Ferdinand's brothers). He could have dedicated this first (and, as it turns out, only) book to Ferdinand in his own name if it had been his own work, or to one of the party at Ciampoli's who had supposedly been so avid to see it.[10] But suppose that the real author were Galileo: the dedication could not be to anyone less than Ferdinand II, yet the rationale would have to be left somewhat abstract so as not to reveal the actual roles of the personalities. A dedication from Aggiunti would have been dishonest if he were not the real author, but a dedication on behalf of the university would have been perfectly correct.

Pieralli wrote the dedication on August 14, 1627, in Rome, and Galileo sent a copy to Kepler on August 28 from Florence. That is to say, he sent it as soon as it was available, as if he had been waiting for it to arrive. Perhaps the letter of recommendation, which is short and vague and seems to have no real purpose, was just a pretext for sending the *Oratio*, and the book was the real reason that Galileo wrote to Kepler. It wasn't really to Kepler's taste, but Galileo knew Kepler as a daring and free thinker, and Galileo was eager to show it to him, just as Kepler had been eager to show Galileo the *Mysterium Cosmographicum* so many years before.

The *Oratio* was written at a time when Galileo was known to be procrastinating on the *Dialogue*, whether for reasons of health or for other reasons, a cause of great concern among all his friends. Perhaps he had turned his attention to the *Oratio* instead, without telling them. It makes a tiny book, just thirty-four pages, but it is not insubstantial either. He might have noticed the one-time opportunity to disguise this speculative work, which was perhaps of genuine importance to him, as Aggiunti's inaugural address. Work on this project might explain why he was neglecting the *Dialogue* just at this time.

Niccolò Aggiunti died young, in 1635, and Pieralli honored him with a memorial oration that is the source for much of what we know about him. He recounts Aggiunti's brilliance as a student, the deep impression he made on all those who heard his doctoral defense in 1622, the eagerness with which Ferdinand II made him Professor of Mathematics in 1626, and the love his students had for him and for his eloquent lectures. He also notes the many unpublished orations, discourses, epigrams, and poems that Aggiunti left behind. Somehow in all of this he manages never to mention Aggiunti's only published work, the one that he himself had seen through the press.[11]

The Geometry of the Senses

One of many characteristically Galilean ideas in the *Oratio* occurs near the beginning, as the Orator makes an unorthodox case for the importance of mathematics.

And to begin with, I have no hesitation in boldly affirming and repeatedly declaring this: that nothing is fitter for a man or more suitable to the sharpest intellect than Geometrical speculation. And I will say more, and even nearer the truth, that all other disciplines besides this one are either nothing, or they proceed from it as from their source. I have no fear that I shall not win you over to this truth—it may seem distasteful, but it is most certain. You may pass in thought through Heaven and Earth, you may journey in imagination through the whole universe of creation,[12] and in all things, of which the appearances reach the mind by way of the senses, what do we have, what do we perceive except motion, color, number, and shape? Really nothing. If, therefore, we wish to have knowledge of anything, it will be the work of the mind to convert it completely into these things, and for our ingenuity then to deal with them precisely. And only Geometry contemplates these things carefully and investigates them thoroughly.[13]

Galileo had said something very similar about the senses in *The Assayer*, with the slight difference that color was not included in the list of real, physical properties: "To excite in us tastes, odors, sounds I believe that nothing is required in external bodies except *shapes, numbers, and slow or rapid movements* [italics added]. I think that if ears, tongues, and noses were removed, shapes and numbers and motions would remain, but not odors or tastes or sounds. The latter, I believe, are nothing more than names when separated from living beings."[14]

In *The Assayer* Galileo is only speculating on where the line should be drawn between external phenomena and our internal representations of them. The *Oratio*'s version of this idea is more explicit than *The Assayer*'s in pointing out the implications for a mathematical theory of nature. The version in the *Oratio* is uniquely clear and significant for understanding Galileo's mathematical physics. In the *Oratio* both the relevant geometry and the redacted sensory data that it deals with are the products of the human mind, and little attention is given to what may be known in any other sense.

It is not difficult to believe that the sense of sight is essentially governed by geometry. That much is just the content of the painter's theory of perspective. But to imagine that all the senses are somehow geometrical—that is a much more daring idea, perhaps a generalization from the geometrical theory of vision. In focusing attention on the senses themselves, and on the way we acquire information about the world, the Orator is implicitly questioning how we know anything at all. It is the kind of radical questioning of basic assumptions that Descartes more famously initiated at about the same time. One could say that the Orator's idea of a mathematics of phenomena is not further developed here, or on the other hand one could say that natural science after Galileo is the further development of this idea, but with less attention to the role of the senses and the human mind than Galileo would have given it. Newtonian physics for many decades was assumed to have an objective significance that only began to be reinterpreted in the twentieth century, in the theories of relativity and quantum mechanics, as having again a subjective component, in which the observer played an essential role in what was to be understood.

The Conditions of Civilization

The Orator tells the story of Aristippus, shipwrecked on the shore of Rhodes, who noticed, to his delight, geometrical diagrams drawn in the sand. "Be of good cheer!" he called to his comrades, "I see the signs of men." And by this he did not mean just footprints, which are only the tracks of animals that walk, but rather geometry, the tracks of animals that think.[15]

And who led humankind to move from fields and cliff-hewn caves to form a more civilized society with gentle customs in settled homes, if not Mathematics, the instructress of Architecture? The Orator gets a lot of mileage out of this idea, not just the history of the architecture of buildings, but the development of neighborhoods, towns, cities, and eventually commerce and shipbuilding.[16] The mechanical contrivances of civilized society are enumerated, not omitting that most wonderful invention, the Archimedes screw, and not forgetting the war machines with which Archimedes defended Syracuse. The Orator names quite a few other Hellenistic inventions from less familiar inventors like Ctesibius and Ctesiphontis, such as those for using water power, turning wheels, and telling time.[17] And for the governance of cities, and for justice among men, mathematics is indispensable, as for example in the arithmetic that makes fair dealing possible in business. Is not Justice herself represented with a mathematical instrument in her hand, the balance?[18]

This much of the oration was devoted to what is useful, the Orator says, but now let us consider what is pleasant and uplifting to the spirit, beginning with the art of painting. Once men have begun to paint, what is more natural to discover than the mathematics of Perspective? The Orator briefly describes how this theory works, something that most Renaissance mathematicians would not have thought about. He describes optical effects in the atmosphere, rainbows, halos, and the twinkling of stars, subjects of Galileo's attention as he made his first telescopic observations and tried to understand the apparent sizes of heavenly bodies. Indeed, these thoughts call to the Orator's mind Galileo's telescope, which is praised to the skies. Galileo is called the Florentine Prometheus and the Etruscan Atlas. It is noteworthy that the telescope is not associated so much with astronomy here as with optics, and even with painting, if we look at the sequence of ideas.[19] As if to emphasize the telescope's nonastronomical meaning, it is immediately followed by an invention of Galileo (he was not the first, nor did he ever claim priority) for which there is very little documentation, his "micro-telescope," as it is called here:

> Is it not a delight that is charming, thrilling, and useful? In the tiny and minute bodies of little animals nature showed most especially her greatest cleverness, but in the past these creatures escaped the dull sight of our eyes, which were too feeble to examine the intricate subtleties of this minute work. Recently, however, with the help of the microscope we have so sharpened our vision that we can look at all the parts of any tiny creature whatever and examine the smallest features limb by limb; with this lynx-sharp eye we can at our leisure examine in insects, whether they fly or live in the ground, their minute hooked or bifurcated feet, their hairy little legs, their scissor-like mouths, their many-colored or varicolored wings, their reticulated eyes— in a word, their entire appearance; and as we examine their whole configuration with minute attention, we are suffused with an intense joy. This joy is indeed extraordinary, subtle, and full of an almost divine quality.[20]

The Orator seems to have spent a lot of time looking with intense curiosity through Galileo's microscope. This is perhaps the fullest description in words of what Galileo saw there, although he published a page of sketches of the magnified bee in 1624.[21] There was never the same public enthusiasm for Galileo's microscopes that there was for his telescopes. For one thing, he made no galvanizing discovery that people wanted to see. For another, it is much harder to make and use a crude microscope than it is a crude telescope. You can hold two appropriate lenses in your hands, one in front of the other, look through them, and get the effect of a telescope, as della Porta had said long before Galileo ever made a telescope, but you cannot easily do the same thing with a microscope. The tolerance for positioning the objective lens is much less forgiving. It has to be fixed firmly in just the right place. It was not easy for amateurs to come into this field.[22]

Although the microscope could be the subject of a very rewarding oration all by itself, he says, many other mathematical arts are summoning the Orator back to his theme, and most especially that siren, the artificer of Music. He speaks knowledgeably and at length about the effects of the musical modes, Dorian, Lydian, Ionian, and Harmatian,[23] and marvels at how an effect that is almost nothing in the physical sense, just a slight perturbation of the air or a tiny vibration of the eardrum, can move us so powerfully and so profoundly. He describes briefly the phenomenon of resonance: strings consonant with a given string (he says) also begin to sound when the given string is played, and perhaps something like that is happening in the soul to explain our response to music. Galileo describes resonance in *Two New Sciences* at considerably more length, but without the psychological speculation.

A number of curious topics follow under the heading of wonders, especially theatrical effects. It is probably just coincidental that in his *Dialogue on Ancient and Modern Music* Vincenzo Galilei had described Roman music degenerating to become merely music of the theater, a progression of associations similar to what the Orator produces here. Mathematics is useful in the control of water, including the design of ornamental fountains in gardens. Hellenistic wonders driven by water power, steam, or compressed air are mentioned, as well as an Egyptian statue that would sing at dawn, and Cleopatra's barge.

Finally Astronomy gets a very few words, calling up navigation, which also gets just a very few words.

The Generality of Mathematics

The Orator has reminded us how mathematics is first necessary, then useful, then delightful, but even more than all this, it has a general character in teaching us to argue and to think effectively. The style of argument here is characteristically Galilean:

> Although mathematics seems on the surface only to put together ideas about triangles, circles, and other shapes,[24] I contend that as we work we learn at the same time how to put together ideas about other things, weigh their importance, and solve a problem by a correct argument. When a lute player learns to play a particular tune on the strings, he achieves not only that one thing that he is especially studying and working on but, as he makes his ear more sensitive and trains his fingers to play quickly and distinctly, he learns how to call forth the notes and subdivide the rhythm in any other tune. Having acquired a good ear and nimble fingering, he is fully capable of producing any tune correctly and playing the lute correctly in any circumstances.
>
> A painter who has worked hard at delineating nothing but human faces and figures nevertheless, if he cares to, will be able to draw Peruvian plants, Nile hippopotamuses, and any African monster, even if he has never seen it before, with perfect accuracy of line. Although until now he has drawn only human forms, he has learned how to depict in two dimensions not only them but also any other three-dimensional thing [Lat. *mole*] in any orientation.
>
> Just so, clearly, the mathematician, as he thinks about his figures and works out a true conclusion, not only constructs complete arguments in geometry and arithmetic, but improves his mind, exercises his intellect, sharpens his mental power, and trains, teaches, and strengthens his thinking so much that he can discourse on any other subject without error, state certainties confidently, leave uncertainties to one side, and with his eyes

well open, in the face of ambiguity, maintain a supple engagement with phenomena [Lat. *lubricum assensum*], so that he is not deceived by falsehood looking like truth, as Ixion was deceived by a cloud looking like Juno.[25]

Apart from the metaphor of the arts for mathematics, which is Galileo's own fingerprint, there are interesting hints here of not just the certainties of mathematics, but of the less-familiar process of using mathematics in a situation that is uncertain, like an experimental situation. This would not have been easy for a contemporary audience to understand, for whom mathematics and certainty were virtually synonymous.

Education and Philosophy

The Orator has firm opinions on education, not inappropriate if the oration was addressed to university students. It is the kind of thing Galileo must have said frequently to his own students. This section foreshadows what Galileo said to Viviani a few years later on the vanity of memorizing the names of logical distinctions.

> All the ancient philosophers admitted no one to the study of philosophy (which is the whole work of the mind and of reason) unless he had first made a thorough study of mathematics. Once upon a time, when a man who had not yet laid down the rudiments of study under the guidance of a mathematician applied to attend the philosopher Xenocrates's school, Xenocrates said to him, "Go away! I don't soak fleeces here."[26]
>
> Nowadays, instead of mathematics, young men are taught

who-knows-what logical quibbles and formulas for making distinctions. They memorize these and use them whenever the opportunity arises, as if they were interchangeable adornments. Prepped and stuffed with these foolish phrases, the lads play the game they have agreed upon in law courts and pulpits, infantilizing philosophy. What is more laughable and infuriating is that great men stand about with their ears pricked up and ordinary folk marvel in respectful silence at these idiocies, which are then lustily cheered. Yet the boy who is twittering from the dais understands what he is saying no more than the lute hears its own music—you might say to him what the man said to the skinny nightingale when he'd plucked out all its feathers: "You're just a voice and nothing else!"[27]

We may rightly wonder if a young professor of mathematics would have challenged the philosophy faculty so bluntly, or at all.

In what would someday be unremarkable but was close to unthinkable in 1627, the Orator flatly contradicts Ptolemy's sharp distinction between philosophical mathematics and earthly physics:

All earthly objects show forth the divine mathematics to those who observe them with close attention. They proclaim with utmost clarity that God is the Archgeometer: the movements of the stars; the balance of the earth; the absorption by plants of moisture from the ground through fibrous pipes, as though through a siphon; the penetration of the moisture to the leaves by means of veins running through the whole trunk and branches; the swimming, flight, and crawling of fishes, birds, and reptiles—obviously a subtle hidden mathematics underlies all these phenomena.[28]

Intellectual Pleasure

The oration's conception of mathematics is peculiar in yet another way, in its emphasis on the intellectual pleasure of mathematics, exceeding all other pleasures. Such a thing as sexual pleasure, for instance, cannot compare with mathematics: what lover, having finally spent the night with the girl of his dreams, has ever spent the next morning sacrificing an ox in gratitude? Yet Pythagoras sacrificed a hundred oxen in thanks to the gods for his most famous theorem, quite a difference.

The Orator makes high claims for mathematics, but they are almost a parody of the usual high claims. His mathematics is an entirely human one, its rewards not abstractly virtuous but rather immediate, and almost sensual, beyond sensual. Mathematics is not about the goodness of the cosmos, but about the excellence of the human mind, its inventiveness, its ability to discern with accuracy and clarity, its pleasure in its success after hard intellectual struggle. Mathematics is an entirely human endeavor. How ironic that Galileo's own story is often told in the older moralizing terms that invested mathematics with objective religious meaning. Even if he became the hero of that story, the terms are those of his opponents, who successfully recast Galileo as something that he was not, inflexible and dogmatic. Their success has made the *Oratio* difficult to recognize as a characteristically Galilean work, and made it difficult to recognize what Galileo's life's work really was.

Proportion

The *Oratio* proceeds entertainingly to its end, but the auditors (or the readers) must have had only the vaguest idea of what all of this had to do with mathematics. The mathematical methods of the arts and other earthly things, the main subject of the oration, formed no part of the mathematics curriculum in any university, and hence would not have been considered mathematics by a contemporary academic audience. Prevailing philosophy had taught this audience that earthly things could not be mathematical in any but the most trivial sense.

The Orator, on the other hand, is pointing out how much mathematical theory there already is for earthly things, whatever philosophy may say, especially in the arts, and how much more we can expect to discover. The Orator ignores Astronomy not because it is irrelevant to mathematics but because the issue does not arise there: everyone knows that the heavens are mathematical. He is arguing for an earthly mathematical philosophy.

It is still difficult, even for us, with a well-worked-out mathematical physics, to understand how the Orator envisions a general mathematical approach to all phenomena, going beyond the particular practices and inventions that he describes in the *Oratio*. He clearly does intend mathematics to encompass all of it in some unifying way. It goes beyond what the *Oratio* says to speculate on this, but it is worth noticing what Galileo would undoubtedly have meant, in large part: the unifying idea is just that notion that is so central in all the arts, the idea of *proportion*.

Galileo's ultimate formulation of the parabola law, to take the best concrete example, involves a proportionality between the vertical speed gained by a falling object and the time of its fall. This proportionality was at the heart of the phenomenon, as he saw it. Experiment had played a role in this formulation and that brought in more proportionalities. A measurement of time itself is impossible. Rather one measures something else that ought to be proportional to time, the water that flows through a water clock, or the oscillations of a pendulum. An experimental determination of speed is even less obvious, but Galileo often thought of speed, constant speed at least, as proportional to distance traveled in fixed time. From Galileo's point of view, which stayed very close to raw sensory information, it was not at all a simple thing to say that a speed was proportional to a time. It was not even something that he could verify directly. Rather he had to transform this statement by mathematics into something more directly observable.

Virtually all of Galileo's discoveries make creative use of the notion of proportionality. His discovery of scaling laws in the near-debacle of the Inferno Lectures, for example, is an insight that derives from proportions, although not the proportions of *concinnitas* but rather physical ones. They come from geometry: under scaling, weight goes as the cube of the scale factor, area goes as the square, strength goes as the area, and so on. It is all about proportions grounded in geometrical reasoning and experience.

When Galileo finally summarized in *Two New Sciences* the results of those experiments in music completed with his father so long ago, he did it in terms of proportionalities, although not the proportions of Pythagorean harmonies but rather physical proportions involving several variables: pitch, length, tension, and weight, an efficient way to state everything that is physically relevant in this problem.[29] He says essentially what modern textbooks say, except that in a modern book this would be considered a theoretical result. Absent the theory, Galileo's statement is an experimental result. It is probably the first example known to him of unexpected proportionalities in nature revealed by experiment, and perhaps in that sense the template for his later experiments on motion, using Pythagoras's "method of philosophy."

The notion of proportion, central in all the arts, took on a new significance in Galileo's work. The existence of unsuspected proportions in nature, waiting to be discovered, became a unifying theme in his thought. Because of unavoidable experimental error one cannot literally discover proportionalities in nature if one does not first assume that they are there. This must therefore have been Galileo's metatheory: there are hidden proportionalities among things that geometrical reasoning can discover or that well-constructed experiments can establish. Once established, these proportionalities, together with their mathematical consequences, become the theory.

The idea of proportion in the arts arose in various contexts, from the geometrical proportions of a constructed figure in a painting to the simple integer ratios of music and the *concinnitas* of architectural theory. A subtle difficulty standing in the way of regarding all such proportions as being abstractly alike had been noticed already by the Pythagoreans in their discovery of irrational proportions in geometry, exactly those proportions that could not be expressed with integers. These subtly different notions of proportion had even become a part of controversies on musical tuning, as Galileo's father had suggested that perhaps an irrational tuning was what performers really used.

The Hellenistic Greeks had recognized the importance and subtlety of defining what it means for pairs of dissimilar things to have the same proportion. Their sophisticated resolution of this problem was not really understood again in modern times until the nineteenth century. The key idea is to be found in Euclid's *Elements* V, Definition 5, on the Eudoxian theory of proportion. In the context of his own mathematical philosophy of phenomena, Galileo recognized Definition 5 as crucial. Archimedes's proof of the law of the lever, for example, a touchstone of Galileo's intellectual world, was actually two proofs, one in case the weights have a rational proportion, and a different one in case they have an irrational proportion. The second proof is just an implementation in a particular physical context of Euclid's Definition 5. We know that this foundational concept was regarded as crucial by Galileo, because he says so very clearly in the dialogue that he was working on at the end of his life,[30] intended as a sequel to *Two New Sciences*. It was precisely to Euclid's Definition 5 that Galileo turned in his last writings, dictated to his amanuensis Viviani. Approaching eighty years of age himself, he brought back his old friends Salviati and Sagredo in all their intellectual vigor, delighted to meet and talk again.

Even Simplicio is eager to see what he can make of the Eudoxian theory of proportion.

Nowadays, with our theory of the real numbers unifying the rational and the irrational, we no longer feel the same qualms about the notion of proportionality that Galileo did, but that is only because the matter has been clarified along the lines that Galileo implicitly suggested. As for the idea of proportion, it remains at the center of mathematical physics. It is a surprisingly concise and apt description of physics to call it the science of proportionalities in nature.

Epilogue

It is not easy to write about Galileo and yet to ignore the Copernican controversy, as I have done until now. In the end I feel a responsibility to record my view, namely that the importance of the Copernican controversy in Galileo's biography is overstated. The archetypal dogmatic Copernican is an image that was constructed for Galileo by other people, it must be said, first by his persecutors, and later, when it appeared that he had been right after all, by those who regarded themselves as his successors. There is every reason to regard it with skepticism, and even to doubt the depth of Galileo's interest in the Copernican question.

The Copernican controversy crowds out important elements of Galileo's story, and distracts attention from what Galileo's scientific legacy really was. In fact, none of his students, even those like Castelli who had done astronomical observations with him, went on to work particularly in astronomy, much less to concern themselves with cosmological questions. They typically worked on problems involving properties of solids and fluids, pressure, temperature, and geometry. Evangelista Torricelli, who was with Galileo at his death, invented the barometer, and is remembered today by a unit of pressure, the torr.

Sagredo invented a thermometer, and Viviani was the first to devise a thermometric scale based on proportion.[1] Castelli worked with Galileo on the phenomenon of floating, and later on practical problems of hydrology. Aggiunti investigated the effects of surface tension in capillarity and other settings, and the behavior of vibrating strings.[2] These are just the sorts of problems that the Orator was describing. In this light, one could argue, Aggiunti's *Oratio* fairly characterizes the kinds of problems that Galileo and all his students worked on, making it not an idiosyncratic work, but rather a representative Galilean document.

Galileo's great editor Antonio Favaro knew all of this, yet he still found Aggiunti's oration so peculiar and baffling that he omitted it from the national edition of Galileo's work. Favaro's judgment has been the last word on the *Oratio* until now, but this is a decision that should be re-examined. The *Oratio* arguably represents Galileo's own view of mathematics and the scientific enterprise, perhaps even in his own words.

I feel I must also make some sense of Galileo's Copernicanism, which I maintain was not as central to his thought as most people have believed. To anyone who believed that Galileo was a committed Copernican, his apparent coyness in never quite saying so would be a continual source of frustration. Kepler in 1610 was sure that Galileo was a Copernican, and his brave and lonely support of Galileo in the first weeks after the announcement of the telescopic discoveries was in part motivated by this affinity that they presumably shared.[3] It is certainly true that Galileo had found the Copernican hypothesis very probable and attractive, as he had avowed in his 1597 letter to Kepler. That letter, though, hinted that his favorite arguments for the motion of the Earth were not astronomical ones but rather terrestrial phenomena, and especially, as Kepler correctly guessed, the "flux and reflux of the sea," the tides.[4] Galileo's own later telescopic discoveries showed him the phases of Venus, proof that Venus orbits the Sun. And the apparent sizes of the outer planets, as seen in the telescope, now larger, now smaller, made it clear that their orbits were more nearly centered on the Sun than on the Earth, a compelling if qualitative fact. The technical discoveries of Kepler, though, are much stronger evidence for Copernicanism, and they never interested Galileo in the slightest. And so far from appreciating the meticulous observations of Tycho that underlay Kepler's discoveries, Galileo seemed to dislike everything connected with Tycho, for reasons that are not so clear.

None of the astronomical arguments for Copernicanism known to Galileo (and that was only a subset of the arguments that he could have known, if he had been serious about it) was conclusive, and as an admirer of Aristotelian logic he could not honestly claim to derive a result that did not follow from the premises. His lifelong caution on Copernicanism was grounded in simple logic: the evidence that he knew was not sufficient. The strongest argument, in his mind, was always the nonastronomical one, the tides, and his great *Dialogue on the Two World Systems* was intended originally to feature the phenomenon of the tides in its title, a subject that still dominates its Fourth Day. The tides seem to show the motion of the Earth quite directly, and Galileo was particularly astonished that Kepler, of all people, someone who really was a committed Copernican, nevertheless did not join him in understanding the tides as driven by the motion of the Earth, rather ascribing the tides to the occult influence of the Moon, a "puerility" that Galileo was at a loss to understand.[5]

When in 1637 Galileo discovered the libration of the Moon, a subtle rocking back and forth of the Moon's face, he also believed that he had found in these motions periods that corresponded to the tides, and with this discovery he quietly accepted that the Moon does govern the tides after all.[6] With the removal of the tides as an argument, and in light of the pronouncements of the Church on the subject, his commitment to Copernicanism, never complete, seems to have suffered notably. Indeed, in a letter of March 1641 he says,

The falsity of the Copernican system must not on any account be doubted, especially by us Catholics, who have the irrefragable authority of Holy Scripture interpreted by the greatest masters in theology, whose agreement renders us certain of the stability of the earth and the mobility of the sun around it. The conjectures of Copernicus and his followers offered to the contrary are all removed by that most sound argument, taken from the omnipotence of God. He being able to do in many, or rather in infinite ways, that which to our view and observation seems to be done in one particular way, we must not pretend to hamper God's hand and tenaciously maintain that in which we may be mistaken. And just as I deem inadequate the Copernican observations and conjectures, so I judge equally, and more, fallacious and erroneous those of Ptolemy, Aristotle, and their followers, when [even] without going beyond the bounds of human reasoning their inconclusiveness can be very easily discovered.[7]

This letter does not fit well into the familiar narrative of Galileo's life, and most Galileo biographies avoid mentioning it or consider it an obvious prevarication. If one takes it seriously, though, it suggests that the usual biographical narrative, unable to accommodate this evidence, must have omitted something important. A careful reader will note that the letter is not inconsistent with the cautious scientific positions that Galileo always assumed, although in deferring to the theologians it expresses that caution in a rather startling new way.

Remarkably, Galileo does not at all give up on the scientific evidence in this letter. He was writing to a long-time correspondent, Francesco Rinuccini, who was puzzled by two observations that he had recently heard about, one seeming to favor Copernicanism and one seeming to negate it. Galileo discusses these ideas, long familiar to him, in the terms that he always used, patiently pointing out why neither observation is at all conclusive. Galileo's letter, surprising as it is, shows him continuing to engage the scientific question in just the way that he always had, only no longer believing that there was evidence to threaten the Church's position.[8] The evidence seemed less compelling to him in 1641 than it had even just a few years earlier. Believing, perhaps, that the Copernican question might never be decided scientifically, he was willing to defer to other authorities.[9] The phrase "especially by us Catholics," as though Catholics might take one view and Protestants another, makes most sense if what is moving and what is stationary is not an objective fact, in the scientific sense. That Galileo might have come to think about it in this way is easier to accept if one understands that Galileo never had a stake in a particular outcome of the Copernican controversy, but that he saw it rather as a problem for the human intellect confronting the evidence available to it, the question of how to philosophize, as he might have said. In the absence of conclusive evidence, though, the Copernican question had become a less interesting problem. The Rinuccini letter, which might seem to be a renunciation of his principles, was perhaps a remarkable example of his adherence to them.

If Galileo was not quite what the standard myth makes of him, we may ask how we should rather understand him, given that he is, for many purposes, not so much a historical person as he is a symbol, the personification of an imagined conflict between science and religion, truth and authority, or almost, to put it baldly, right and wrong. Such a caricature is a misrepresentation of him, not just mistaken but pernicious in its widespread acceptance, yet it does little good to argue the fine points of the case in a scholarly way. A careful argument will never prevail over a simple and instantly understood caricature. Galileo has an image problem. Galileo needs a public relations makeover.

Perhaps what is wanted is a new caricature, one more useful and nearer the truth. In this way we re-enter, having left them on the title page, the territory of the Muses, those slightly cartoonish but enduring divinities, almost the last of "Olympus' faded hierarchy" to have any currency with us. They are ancient daughters of Zeus, invoked by Homer and named by Hesiod, and in later times variously associated with specific arts, or even subjects we would not immediately call arts, like Clio, Muse of History, or Urania, Muse of Astronomy. Urania, I am arguing, is not Galileo's Muse, but she is close enough to indicate what is missing among the Nine, a new art, requiring a new Muse who really would be his, something like a Muse of Earthly Things, or a Muse of Mathematical Experimental Science. If Astronomy could have a muse in Hellenistic times, then surely these more modern subjects, so aptly associated with Galileo, could have one now. There are many practicing scientists who would be glad to imagine that this Muse was hovering nearby to symbolize to others, or even to themselves, what it is that they do. She would need a name—Galilea? Such a divinity would personify Galileo's notion of science with about the right mix of seriousness and lightness, I think. It would take a long time to establish her, probably as long as it has taken to establish the old Galileo story. I can imagine a distant future time, though, when it would be common knowledge, and simply assumed, that Galileo, sometime around the beginnings of our present scientific age, performed a Pygmalion-like trick: he invented a new Muse to smile on him.

Notes

1. Galileo, Humanist

1. Galileo's extensive correspondence with potential publishers, leading to the eventual publication, is detailed in Stillman Drake, *Galileo at Work* (New York: Dover, 1995), 365–384.

2. R. H. Naylor, "Galileo's Experimental Discourse," in *The Uses of Experiment*, ed. D. Gooding, T. Pinch, and S. Schaffer (Cambridge: Cambridge University Press, 1989), 117.

3. Galileo, *Two New Sciences*, tr. Henry Crew and Alfonso de Salvio with an introduction by Antonio Favaro (New York: Dover, 1954), 242 (end of the Third Day).

4. Consider the motivation stated in Galileo's previous book, the great *Dialogue Concerning the Two Chief World Systems* (1632), the work for which he was hauled to Rome for trial: was it really true, as he claims there, that he wrote the *Dialogue* to demonstrate that the Church had understood all the arguments in favor of Copernicanism in 1616, and had therefore banned Copernicanism *wisely?* Even at the time no one thought that this was his real opinion, and the Inquisition experts jumped all over this artifice.

5. See essays on Galileo and music by Claude Palisca and Stillman Drake in *Music and Science in the Age of Galileo*, ed. V. Coelho (Dordrecht: Kluwer Academic Publishers, 1992) and Stillman Drake's essays in *Essays on Galileo and the History and Philosophy of Science*, vol. 3 (Toronto: University of Toronto Press, 1999), 190–220.

6. E. Panofsky, *Galileo as a Critic of the Arts* (The Hague: Martinus Nijhoff, 1954).

7. Galileo, *Le Opere di Galileo Galilei I–XX*, ed. Antonio Favaro, ristampa della edizione nazionale (Florence: G. Barbèra, 1929–1939), XIX. For Viviani, see 599–624; for Gherardini, see 634–646; for Vincenzo, see 594–596.

8. Viviani mentions the trip to Rome only to say that it was the occasion of Galileo's writing a long, privately circulated treatise on the tides, at the request of Cardinal Orsini, based on the motion of the Earth.

9. From *The Assayer*, in *Discoveries and Opinions of Galileo*, tr. and ed. Stillman Drake (New York: Doubleday Anchor Books, 1957), 237.

10. How it might be permissible to use these sources has been considered by Michael Segre, "Viviani's Life of Galileo," *Isis* 80 (1989): 207–231.

11. Mario Biagioli, *Galileo Courtier* (Chicago: University of Chicago Press, 1993).

12. *Galileo Opere* XIX, 596.

13. Galileo to Liceti, 25 August 1640, *Galileo Opere* XVIII, 234, quoted in Drake, *Galileo at Work*, 407–408.

14. The Inquisition's expert Melchior Inchofer wrote in his report, "he writes in Italian, certainly not to extend the hand to foreigners or other learned men, but rather to entice to that [Copernican] view common people in whom errors very easily take root." Quoted in Maurice Finocchiaro, *The Galileo Affair* (Berkeley: University of California Press, 1989), 266.

15. *Galileo Opere* XIX, 643.

16. A. Favaro in *Galileo Opere* IX, 12.

17. *Galileo Opere* IX, 61–148.

18. A recent comprehensive study is Horst Bredekamp, *Galilei der Künstler: Der Mond, die Sonne, die Hand* (Berlin: Akademie Verlag, 2009).

19. Regarding the art scene in Rome, see Cigoli to Galileo, 9 April 1609, *Galileo Opere* X, 241–242; 24 October 1610, *Galileo Opere* X, 456; 13 December 1610, *Galileo Opere* X, 475. Regarding the literary request, see Cigoli to Galileo, 22 May 1609, *Galileo Opere* X, 243–244. Cigoli also became an enthusiastic collaborator in the observations of sunspots in 1611.

20. Gentileschi to Galileo, 9 October 1635, *Galileo Opere* XVI, 318–319.

21. *Galileo Opere* XIX, 602–603.

22. Ibid., 601.

23. Ibid., 604.

24. Gherardini's first acquaintance with Galileo in 1633, during the trial, was

in the capacity of a secret emissary for a high-ranking churchman on the tribunal, offering Galileo advice on the conduct of his defense.

25. *Galileo Opere* XIX, 637.

26. Special Commission Report, point vi, quoted in Finocchiaro, *The Galileo Affair*, 222.

27. Galileo, *Dialogue Concerning the Two Chief World Systems*, tr. and ed. Stillman Drake, foreword by Albert Einstein, 2nd rev. ed. (Berkeley: University of California Press, 1967), 103.

28. Salviati refers to the Pythagoras theorem.

29. *Dialogue*, 104–105.

30. Ibid., 406.

31. Quoted by Stillman Drake from Galileo's notes on a book of Lagalla, *Opere* III: I, 395–396, in *Discoveries and Opinions*, 224–225.

32. John Aubrey, *A Brief Life of Thomas Hobbes, 1588–1670*, in *Seventeenth-Century Prose and Poetry*, 2nd ed., selected and edited by A. M. Witherspoon and F. Warnke (New York: Harcourt Brace Jovanovich, 1963), 495.

33. From the *Discourse of the Sons of Mūsā, Ibn Shākir: Muhammad, Ahmad, and Hasan,* in Marshall Clagett, *Archimedes in the Middle Ages,* vol. I (Madison: University of Wisconsin Press, 1964), 239.

34. *Galileo Opere* III:I, 253.

35. *Galileo Opere* II, 369. Galileo has actually adapted to Archimedes a story that is more often associated with Euclid.

36. Ibid., 307–334, translated as *Dialogue Concerning the New Star by Cecco di Ronchitti,* in Stillman Drake, *Galileo against the Philosophers* (Los Angeles: Zeitlin and Ver Brugge, 1976), 36–51.

37. One of the professors of philosophy was Cesare Cremonini, who is widely thought to have been the model for Simplicio in Galileo's later dialogues, where he is a valued and amiable participant, much as Cremonini seems to have been a colleague whom Galileo valued despite their differences. The new star was the occasion of some public lectures by Galileo on parallax, but the pseudonymous form of the tract gave him much more freedom.

38. Consultants' Report on Copernicanism, 24 February 1616, in Finocchiaro, *The Galileo Affair*, 146.

39. *Galileo Opere* V, 201, translated in Drake, *Discoveries and Opinions*, 135.

40. Galileo to Liceti, 25 August 1640, *Galileo Opere* XVIII, 234, quoted in Drake, *Galileo at Work*, 407–408.

41. *Galileo Opere* XIX, 645.

2. The Classical Legacy

1. A recent comprehensive study, Christoph Riedweg, *Pythagoras* (Ithaca, NY: Cornell University Press, 2005), brings the fragmentary evidence together. The surprisingly lively questions that still surround Pythagoras today are not as relevant for us as the views that Galileo apparently held, relying largely on the same evidence that we have.

2. Archytas "served seven terms as his hometown's official military commander without suffering a single defeat"; ibid., 111. He seems to have intervened on Plato's behalf in a dispute at Syracuse.

3. Ibid., 81.

4. Hippol. *Phil.* 2; *Dox.*, 555, quoted in *The Pythagorean Sourcebook and Library*, ed. K. S. Guthrie (Grand Rapids, MI: Phanes Press, 1987), 312.

5. Galileo, *Dialogue Concerning the Two Chief World Systems*, tr. and ed. Stillman Drake, foreword by Albert Einstein, 2nd rev. ed. (Berkeley: University of California Press, 1967), 11, and accompanying note. The tactic of throwing the inquisitive some foolishness that will satisfy them is illustrated by Galileo with a story that he doesn't actually tell. He merely alludes to it, assuming that his readers will recall a story from Macrobius's *Saturnalia* about the Roman senator Papirius. Papirius's mother was eager to know what the secret debates in the Senate had been about, so he told her that they were debating whether it was better to allow men to have two wives or women to have two husbands. The senators were astonished the next day by a demonstration of townswomen in favor of the second alternative.

6. Hermann S. Schibli, Archytas, in *Routledge Encyclopedia of Philosophy*, ed. E. Craig (London: Routledge, 1998).

7. Lucio Russo, *The Forgotten Revolution* (New York: Springer-Verlag, 2003).

8. For what is known, see T. L. Heath's introduction to Euclid, *Elements*, vol. I, tr. T. L. Heath (New York: Dover Books, 1956), 1–6.

9. Ibid., 98–99.

10. Euclid, *Elements*, vol. 2, tr. T. L. Heath (New York: Dover Books, 1956), 120–129.

11. Euclid, *Opera Omnia*, ed. I. L. Heiberg and H. Menge, vol. 7, *Euclidis Catoptrica* (Leipzig: B. G. Teubner, 1895), 341, Prop. 30: *"e speculis concavis adversus solem conversis ignuis adcenditur."*

12. Plutarch, *Lives*, V, Marcellus, Loeb Classical Library V (Cambridge, MA: Harvard University Press, 1917).

13. Galen, *De Temperamentis* III, 2 Galeni opera ex octava Juntarum editione (Venice, 1609) I. 23, cited in E. J. Dijksterhuis, *Archimedes* (Princeton, NJ: Princeton University Press, 1987), 28.

14. Theon of Alexandria and Apuleius of Madaura. See Dijksterhuis, *Archimedes*, 48.

15. Apollonius, *Conics*, tr. R. C. Taliaferro (Santa Fe: Green Lion Press, 1998).

16. T. Freeth et al., "Decoding the Ancient Greek Astronomical Calculator Known as the Antikythera Mechanism," *Nature* 444 (2006): 587–591.

17. Derek J. de Solla Price, "An Ancient Greek Computer," *Scientific American* (June 1959): 60–67.

18. Archimedes had made a working model planetarium of the motions of the heavenly bodies according to Cicero, *De re publica* I, 14. Compare Cicero, *Tusculanae disputationes* I, 25; and *De natura deorum* II, 34, cited in Dijksterhuis, *Archimedes*, 23. This has always been taken to be a model in which the bodies are spheres moving around in space. It is possible that without the more abstract device in front of him, Cicero just assumed the nature of the planetarium, but a lost work of Archimedes, *On Sphere Making*, has been assumed also to refer to the planetarium, which would therefore have been spheres and not gears.

19. Plutarch, *Lives*, V, Marcellus, 481.

20. Plutarch, *Concerning the Face which Appears in the Orb of the Moon*, tr. H. Cherniss, in *Moralia* XII (Cambridge, MA: Harvard University Press, 1927), 59.

21. Ptolemy, *The Almagest*, tr. R. C. Taliaferro (Encyclopedia Britannica, 1952), 6.

22. Copernicus looked for such sources, and was aware of Pythagorean theories of a moving earth, as he tells us in the dedication to Pope Paul III, citing in particular the Pythagorean Hicetas of Syracuse, whom he calls Nicetas. He succeeded to the extent that Copernicanism came to be called by some "the Pythagorean opinion." Thus, confusingly, in the sixteenth century "Pythagorean" took on the potential meaning of "Copernican." This was never how Galileo used the term "Pythagorean," however. On this point he refers in his *Dialogue* to "Pythagoras (or whoever it was)" as one source of the belief in the Earth's motion, a dismissive way of pointing out that it was actually not Pythagoras. Galileo, *Dialogue*, 188.

23. Ptolemy, *Almagest*, 5.

24. Ibid.

25. This philosophical tenet, central to the Aristotelian picture of the world, was an embarrassment again and again in Galileo's own lifetime. The new star of 1604, the sunspots discovered around 1612, and the comets of 1618 repeatedly contradicted this axiom of an immutable heaven, which was still taken seriously at the time.

26. Plato, *The Republic* 617b, tr. Francis Cornford (Oxford: Oxford University Press, 1945), 354.

27. Aristotle, *On the Heavens* II, 9.

28. K. S. Guthrie, ed., *The Pythagorean Sourcebook and Library*, 301.

29. Cicero, *Scipio's Dream*, in *De re publica*, VI, 18 (Cambridge, MA: Harvard University Press, 1928), 273.

30. Vitruvius, *Ten Books of Architecture*, Book I, chap. vi, 9.

31. Ibid., Book IX, introduction, 9–12.

32. Lucretius, *De rerum natura*, Book V, 656–662.

Poetry

1. A. Speiser, *Klassische Stücke der Mathematik* (Zürich: Verlag Orell Füselli, 1925); J. Callahan, "The Curvature of Space in a Finite Universe," *Scientific American* 235 (August 1976): 90–100; M. A. Peterson, "Dante and the 3-sphere," *American Journal of Physics* 47 (1979): 1031–1035; Robert Osserman, *Poetry of the Universe* (Garden City, NY: Doubleday, 1995); M. A. Peterson, "The Geometry of Paradise," *Mathematical Intelligencer* 30, 4 (2008): 14–19.

3. The Plan of Heaven

1. Thomas Aquinas, *Summa Theologica* I, q. 102, art. 1.

2. Aristotle actually left it to astronomers to determine the exact order.

3. Dante, *Paradiso*, XIII, 101–102.

4. Ibid., XVII, 15–17.

5. A. Speiser, *Klassische Stücke der Mathematik* (Zürich: Verlag Orell Füselli, 1925); J. Callahan, "The Curvature of Space in a Finite Universe," *Scientific American* 235 (August 1976): 90–100; M. A. Peterson, "Dante and the 3-sphere," *American Journal of Physics* 47 (1979): 1031–1035; Robert Osserman, *Poetry of the Universe* (Garden City, NY: Doubleday, 1995).

6. Dante, *Paradiso*, tr. Dorothy Sayers and Barbara Reynolds (Middlesex, England: Penguin Books, 1962), XXII, 124–129.

7. Ibid., XXVIII, 58–60.

8. Dante, *Paradiso*, tr. with notes by R. Hollander and J. Hollander (New York: Random House, 2007), note on XXII, 67.

9. Dante, *Paradiso*, tr. Sayers and Reynolds, IV, 37–42.

4. The Vision of God

1. M. A. Peterson, "The Geometry of Paradise," *Mathematical Intelligencer* 30, 4 (2008): 14–19.

2. Dante, *The Banquet*, II.13.

3. I do not mean to downplay the universe of ideas in *The Divine Comedy*, only to suggest that once Dante had glimpsed the complete structure mathematically, perhaps even in a flash of insight that he later recalled autobiographically, he might have felt impelled to begin.

4. Over seventy Dante commentaries can be searched online at dante.dartmouth.edu.

Painting

1. Draft of the dedication to Pirckheimer for *Four Books of Human Proportions*, MS in the British Museum, quoted in *The Writings of Albrecht Dürer*, tr. and ed. William Martin Conway (New York: Philosophical Library, 1958), 253.

2. Ibid., 58.

3. Vitruvius, *Ten Books on Architecture*, Book 7, preamble, paragraph 11.

5. The Power of the Lines

1. J. V. Field, *The Invention of Infinity* (Oxford: Oxford University Press, 1997).

2. Harry Edward Burton, "The Optics of Euclid," *Journal of the Optical Society of America* 35 (1945): 357–372.

3. P. Grendler, "What Piero Learned in School: Fifteenth Century Vernacular Education," in *Monarca della Pittura: Piero della Francesca and His Legacy*, ed. M. A. Lavin, Studies in the History of Art, no. 48 (Washington, DC: Center for the Advanced Study of the Visual Arts, National Gallery of Art, Symposium Papers XXVIII, 1995).

4. The question of Piero's access to sources has been re-opened with the recent discovery of a manuscript Archimedes that Piero apparently made for himself from the Urbino volume, possibly even before it arrived in Urbino. See

J. Banker, "A Manuscript of the Works of Archimedes in the Hand of Piero della Francesca," *Burlington Magazine* 147, 3 (2005): 165–169.

5. W. G. and E. Waters, *The Vespasiano Memoirs* (London: George Routledge and Sons, 1926), 102.

6. Martin Kemp, "Piero and the Idiots," in *Monarca della Pittura*, ed. Lavin.

7. See also J. V. Field, *A Mathematician's Art* (New Haven, CT: Yale University Press, 2005), ch. 5.

8. Piero's proof says $BC : DE = AB : AD = IC : IJ = CG : JK$. Since $BC = CG$, we have $DE = JK$.

9. See also Field, *A Mathematician's Art*, 162–173.

10. G. Vasari, "Life of Paolo Uccello," in *Lives of the Painters*.

11. See also J. V. Field, "Piero della Francesca's Treatment of Edge Distortion," *Journal of the Warburg and Courtauld Institutes* 49 (1986): 66–90.

12. Martin Kemp, "Piero and the Idiots," in *Monarca della Pittura*, ed. Lavin.

13. Field, *A Mathematician's Art*.

14. Carlo Ginzburg, *The Enigma of Piero* (London: Verso, 2002).

6. The Skin of the Lion

1. Luca Pacioli, *Summa de arithmetica, geometria, proportioni, et proportionalita* (1494, repr. Venice: Paganino de Paganini, 1523).

2. G. Mancini, *L'opera "De corporibus regularibus" di Pietro Franceschi detto della Francesca, usurpata da fra Luca Pacioli*, in *Memorie Della R. Accademia Dei Lincei*, 1915 (Rome: Tipografia della R. Accademia dei Lincei, 1916). Includes an edited edition of Piero's *De corporibus regularibus*.

3. Quoted in *Leonardo Pisano Fibonacci: The Book of Squares*, an annotated translation by L. E. Sigler (New York: Academic Press, 1987), xvi.

4. Tobias Dantzig, *Number: The Language of Science* (New York: Macmillan, 1954), 26, quoted in Frank J. Swetz, *Capitalism and Arithmetic* (La Salle, IL: Open Court, 1987), 13.

5. J. J. Sylvester, "On Staudt's Theorems Concerning the Contents of Polygons and Polyhedrons, with a Note on a New and Resembling Class of Theorems," *Philosophical Magazine* 4 (1852): 335–345.

6. Mancini in *L'opera "De corporibus regularibus,"* 454.

7. Euclid, *Elements*, a cura di L. Pacioli (Venice: Paganino de Paganini, 1509), 30.

8. "Difficillima est proportio; haec est illa quae sola intima altissimae individuaeque trinitatis penetrat."

9. "[Q]uam iustus iudex vivorum et mortuorum olim humano generi retribuet merita ac demerita omnium adinvicem proportionando ut ex sacris aperte elicitur litteris."

10. Pacioli had issued his edition of what was by then the standard translation of Euclid's *Elements* in indignant reply to a new translation from the Greek by Zamberti that had finally gotten Definition 5 right.

11. Mario Biagioli, "The Social Status of Italian Mathematicians, 1450–1600," *History of Science* 27 (1989): 91–95.

12. Kenneth Clark, *Leonardo da Vinci*, rev. ed., with an introduction by Martin Kemp (New York: Viking, 1988), 251.

13. V. Foley, "Leonardo and the Invention of the Wheelock," *Scientific American* 278 (January 1998): 96–100.

14. *The Notebooks of Leonardo da Vinci* (Oxford: Oxford University Press, 1952; reissued as a World's Classics Paperback, 1980), 295.

Music

1. Iamblichus, *The Life of Pythagoras*, ch. 25.

2. Ibid., ch. 15.

3. Plato, *Republic* 401e, tr. Francis Cornford (Oxford: Oxford University Press, 1945), 90.

4. Plato, *Republic* 399a–b, 87.

7. The Orphic Mystery

1. V. Galilei, *Dialogue on Ancient and Modern Music*, tr. C. Palisca (New Haven, CT: Yale University Press, 2003).

2. M. L. West, *Ancient Greek Music* (Oxford: Oxford University Press, 1992), 7.

3. Ibid., 384.

4. Galilei, *Dialogue on Ancient and Modern Music*, 83. The Alexander incident is from Plutarch's *Second Oration Concerning the Fortune or Virtue of Alexander the Great*.

5. Galilei, 4–5.

6. *Orestes*, Vienna papyrus G2315; *Iphigenia at Aulis*, Leyden papyrus inv. 510.

7. Aristoxenus, *Harmonics*, tr. Henry S. Macran (Oxford: Clarendon Press, 1902), 188–189.

8. Plato, *The Republic* 531a–b, tr. Francis Cornford (Oxford: Oxford University Press, 1945), 249–250.

9. Aristophanes, *The Clouds*, tr. William Arrowsmith (New York: Mentor, 1962), 26–27.

8. Kepler and the Music of the Spheres

1. Johannes Kepler, *The Harmony of the World*, tr. E. J. Aiton, A. M. Duncan, and J. V. Field (Philadelphia: American Philosophical Society, 1997), 390–391.

2. Galileo to Micanzio, 19 November 1634, *Galileo Opere* XVI, 162.

3. Galileo to Kepler, 4 August 1597, *Galileo Opere* X, 67–68.

4. Kepler to Galileo, 13 October 1597, *Galileo Opere* X, 69–71.

5. Even by 1610, when he wrote *Sidereus Nuncius*, Galileo had written only four little books: his manual for the geometric and military compass, a second book accusing his former student Baldessar Capra of plagiarizing his first book, and two pseudonymous attacks on local philosophers—not a very distinguished *oeuvre*. Even in retirement at Arcetri he seems to have felt a little self-conscious about this, because he told Gherardini that he had freely given away his lecture notes to students, whence some of them had been published under other names.

6. Johannes Kepler, *Mysterium Cosmographicum*, tr. A. M. Duncan (Norwalk, CT: Abaris Books, 1999).

7. Kepler had probably never heard of Galileo, and had only given the books to Hamberger to pass along to some suitable recipient in Italy. There was no accompanying letter to Galileo, as there should have been if they were intended for him. Hamberger seems to have off-loaded the books at the last possible moment, so as not to have to carry them back.

8. Galileo's disinterest in making astronomical observations led to a truly awkward moment some years later when he missed one of the most significant astronomical events of the millennium, the "new star" of 1604, first observed in Padua by his former student Baldessar Capra, and only learned of it from a Venetian friend many days later. In his defense he wrote that Capra, as the first observer, "should be held in appropriately high esteem, although one wonders whether those who aspire to some noble degree of glory in the mathematical sciences should spend every night of their lives in vigilant observation over the roofs of their houses to see if a new star might appear, so that others, who might get some advantage, should not have the triumph of such a glorious discovery." *Galileo Opere* II, 520.

9. E. Aiton, in Introduction to *Mysterium Cosmographicum*, 19–21.

10. In Kepler's own words from the Dedicatory Epistle to the second edition of *Mysterium Cosmographicum*, "almost every book on astronomy which I have published since that time could be referred to one or another of the important chapters set out in this little book, and would contain either an illustration or a completion of it." Of course by this time Kepler had changed his mind about many things as a consequence of the momentous discoveries he had made, but not perhaps as much as one might expect. On the role of the Platonic solids, for example, he had decided that they were only a part of the story: "in its final and most finished state the proportions of the spheres belong to both the solids and the harmonies in common" (note 1 to chapter 16).

11. Kitty Ferguson, *Tycho and Kepler* (New York: Walker, 2002), 200–204. Only the four largest instruments were left behind.

12. Ibid., 290–294. Kepler surrendered all the data except those that interested him most, the data for Mars, hoping that Tengnagel would not notice. After a few months, though, Tengnagel did notice, and demanded these too.

13. Kepler, *Optics*, tr. William H. Donahue (Santa Fe: Green Lion Press, 2000).

14. G. B. della Porta, *Magiae Naturalis* (Naples, 1589), is largely concerned with using the secrets of nature for the purpose of amazing or deceiving people, as if by magic: things like secret writing, magnets, and cosmetics. Della Porta's suggested use of the telescope is that you could set it up in your bedroom looking out the window to see things at a distance without being seen yourself.

15. Page 202 in Kepler's original 1604 *Optics*, p. 218 in W. H. Donahue's translation.

16. Johannes Kepler, *Kepler's Conversation with Galileo's Sidereal Messenger*, tr. E. Rosen (New York: Johnson Reprint, 1965), 13.

17. Johannes Kepler, *New Astronomy*, tr. William H. Donahue (Cambridge: Cambridge University Press, 1992); Kepler's account is not as disingenuous as it appears: see James Voelkel, *The Composition of Kepler's Astronomia Nova* (Princeton, NJ: Princeton University Press, 2001).

18. Kepler, *Conversation*, 9.

19. In the case of Hasdale, the contact must still have been indirect, because Hasdale became acquainted with Kepler only when Galileo's *Sidereus Nuncius* arrived. In his first conversation with Kepler, however, Kepler confirmed that he had the highest opinion of Galileo. It is odd that Galileo had apparently not encouraged Hasdale to look up Kepler before.

20. Kepler, *Conversation*, 47.

21. In a remarkable coincidence it turns out that there really are two moons around Mars, not discovered until the nineteenth century. At the time it would have been awkward for Galileo to have discovered any new moons, because the French royal family had let him know that they too would like to have new planets dedicated to them, while Galileo was reassuring the Medici that the honor he had bestowed upon them was unique, and would never be duplicated.

22. Mario Biagioli, *Galileo Courtier* (Chicago: University of Chicago Press, 1993).

23. Johannes Kepler, *Dissertatio cum Nuncio Sidereo*, in *Galileo Opere* III: 1, 97–126.

24. Galileo to Kepler, 19 August 1610, *Galileo Opere* X, 421.

25. Kepler, *Conversation*, 28.

26. Horky rebuked in letters, see Kepler to Galileo, 9 August 1610, *Galileo Opere* X, 413–417; Horky in Prague, see Kepler to Galileo, 25 October 1610, *Galileo Opere* X, 457–459.

27. Horky to Sizzi, June 1610, *Galileo Opere* X, 386–387.

28. Galileo to Kepler, 19 August 1610, *Galileo Opere* X, 421–423.

29. Magini to Kepler, 26 May 1610, *Galileo Opere* X, 359.

30. Martinus Horky, *Brevissima peregrinatio contra Nuncium Sidereum* in *Galileo Opere* III: 1, 127–145.

31. Kepler to Galileo, 9 August 1610, *Galileo Opere* X, 413–417.

32. Galileo to Kepler, 19 August 1610, *Galileo Opere* X, 421–423.

33. Horky to Sizzi, June 1610, *Galileo Opere* X, 386–387.

34. Hasdale to Galileo, 31 May 1610, *Galileo Opere* X, 365–367. Hasdale to Galileo, 5 July 1610, *Galileo Opere* X, 390–391.

35. Magini to Santini, 22 June 1610, *Galileo Opere* X, 377; Magini to Santini, June 1610, 378–379; Roffeni to Galileo, 29 June 1610, 384; Cittadini to Galileo, 3 July 1610, 389; Roffeni to Galileo, 6 July 1610, 391.

36. Hasdale to Galileo, 12 July 1610, *Galileo Opere* X, 401.

37. Mario Biagioli, in *Galileo's Instruments of Credit* (Chicago: University of Chicago Press, 2006), chapter 2, argues that Galileo deliberately slowed the process by which other astronomers obtained instruments that could compete with his.

38. Giuliano de Medici to Galileo, 19 July 1610, *Galileo Opere* X, 403–404.

39. "Non ardisco domandarle uno de' suoi libretti, non havendo con lei alcuno merito," postscript, Hasdale to Galileo, 28 April 1610, *Galileo Opere* X, 346.

40. Kepler to Galileo, December 1610, *Galileo Opere* X, 506–508.

41. Kepler to Galileo, 9 January 1611, *Galileo Opere* XI, 15–17.

42. Two of Kepler's guesses involved a red spot rotating on Jupiter. Astonishingly, there *is* a red spot rotating on Jupiter, but Galileo had not discovered it. The discovery concealed in the anagram was the phases of Venus.

43. Kepler, *Conversation*, 14.

44. Kepler never referred to Galileo by his new title, Philosopher, referring to him always as "mathematicus Florentinus." See Massimo Bucciantine, *Galileo e Keplero: filosofia, cosmologia e teologia nell'Eta' della Controriforma* (Torino: Giulio Einaudi, 2003), 193.

45. Johannes Kepler, *De Nive Sexangula*, in a modern edition: *The Six-Cornered Snowflake*, tr. L. L. Whyte (Oxford: Oxford University Press, 1966).

46. Daniel Burckhardt, "Nova Stereometria Doliorum Vinariorum: Zur Fassrechnung Johannes Keplers," available at www.sur-gmbh.ch/private/burki/Studies/kepler_stereometria.pdf.

47. What Kepler discovered, as we might say now, is that the Austrian cask shape, within a family of shapes that Kepler could compute, namely two truncated right circular cones sharing a common base to represent a cask that bulges in the middle, is very close to representing a critical point (a saddle point, in fact) of the volume function at a fixed reading of the measuring rod. The critical shape is a cylinder with length-to-diameter ratio $\sqrt{2}$, and Austrian casks (unlike, say, Rhineland casks) have roughly these dimensions. Thus the volume of the Austrian cask, at fixed value of the measuring rod, is surprisingly insensitive to changes in the shape, and is very well measured by the method that Kepler had found so dubious. Kepler painstakingly determined the behavior of the volume function in the vicinity of the critical point. His fascination with this problem seems justified when we consider that the behavior of functions near (singular) critical points continues to be a lively research area in algebraic geometry to this day.

48. Kepler, *Harmony*, 253.

49. Owen Gingerich, "Kepler, Galilei, and the Harmony of the World," in *Music and Science in the Age of Galileo*, ed. V. Coelho (Dordrecht: Kluwer, 1992), 45–63.

50. In *Galileo as a Critic of the Arts* (The Hague: M. Nijhoff, 1954), Erwin Panofsky suggests that Galileo rejected Kepler's First Law for essentially aesthetic reasons, preferring classical circles to mannerist ellipses, but this line of argument has nothing to say about why he also rejected Kepler's Third Law.

51. Kepler pointed out that Jupiter's moons obey his "harmonic law" in his *Epitome of Copernican Astronomy*, 1620, but without giving the straightforward ar-

gument from data. In fact, Kepler didn't demonstrate clearly the accuracy of the law in its original formulation for the planets either, giving a surprisingly vague example in *Harmony of the World* that used only approximate data. Newton made a very careful comparison with data for the moons of Jupiter in the *Principia*, where the "harmonic law" is strong evidence for the universal law of gravitation.

Architecture

1. Antonio Manetti, *The Life of Brunelleschi*, notes and critical text ed. Howard Saalman, tr. Catherine Enggass (University Park: Pennsylvania State University Press, 1970).

2. Leon Battista Alberti, *On the Art of Building*, introductory essay by Joseph Rykwert (Cambridge, MA: MIT Press, 1988), xii.

9. Figure and Form

1. Leon Battista Alberti, *On the Art of Building*, tr. Joseph Rykwert (Cambridge, MA: MIT Press, 1988), Book 9, Ch. 5, 301–302.

2. Quoted in R. Wittkower, *Architectural Principles in the Age of Humanism* (London: Warburg Institute, 1949), 103.

3. Ibid., 100, footnote 1.

4. Alberti, *On the Art of Building*, Book 7, Ch. 10, 219.

5. Vitruvius Pollio, *I dieci libri dell'architettura di M. Vitrvvio, tradotti et commentati da monsig. Daniel Barbaro* (Venice: Francesco de'Franceschi Senese, 1584).

6. Alberti, *On the Art of Building*, Book 9, Ch. 7, 310.

7. Ibid., Ch. 8, 311.

8. Copernicus, *On the Revolutions of the Heavenly Spheres*, in *Great Books of the Western World*, vol. 16 (Encyclopedia Britannica, 1952), 507.

9. Alberti, *On the Art of Building*, Book 4, Ch. 3, 102; Vitruvius, *I dieci libri dell'architettura di M. Vitrvvio*, Book 1, Ch. 5.

10. Horst de la Croix, "Military Architecture and the Radial City Plan in Sixteenth Century Italy," *Art Bulletin* 42 (1960): 263–290.

10. The Dimensions of Hell

1. M. A. Peterson, "Galileo's Discovery of Scaling Laws," *American Journal of Physics* 70 (2002): 575–580.

2. This point is emphasized by Mario Biagioli, *Galileo Courtier* (Chicago: University of Chicago Press, 1993), 106–107.

3. The text of the lectures is in *Galileo Opere* IX, 31–57.

4. Ibid., 7.

5. Ibid., 33–34.

6. Galileo refers to Archimedes's *On the Sphere and the Cylinder.* The relevant propositions are 42 and 44 of Book I. The precise volume fraction is $(2 - \sqrt{3})/4$, a number between $\frac{1}{14}$ and $\frac{1}{15}$, as Galileo says.

7. *Galileo Opere* IX, 48–49.

8. In *The Four Books on Human Proportion,* Dürer constructs human figures seven, eight, nine, and ten heads high. See Dürer, *Writings,* tr. and ed. William Martin Conway (New York: Philosophical Library, 1958), 235–237.

9. *Galileo Opere* IX, 51–52.

10. Ibid., 53–54.

11. Ibid., 54–55.

12. Ibid., 56.

13. Ibid.

14. Ibid., 56–57.

15. Ibid., 57.

16. Galileo, *Two New Sciences,* tr. Henry Crew and Alfonso de Salvio (New York: Dover, 1954), 3.

17. *Galileo Opere* IX, 7.

18. Ibid., 8.

19. Galileo, *Two New Sciences,* 131.

20. "Non si può compartir quanto sia lungo, / Si smisuratamente è tutto grosso," Ariosto, *Orlando Furioso,* XVII, 30.

21. There does seem to have been a tradition of assigning sizes to stars, but it was outside of academic astronomy, i.e., mathematics. See, for example, Leonardo da Vinci, *Codex Urbinas* 15v–16r, cited in *Leonardo on Painting,* ed. Martin Kemp and Margaret Walker (New Haven: Yale University Press, 1989), p. 21, and Daniel Santbech (Basel: Problemata Astronomicorum et Geometricorum, 1561), Prop. XXIII, p. 77.

22. He makes the 1,750 miles at the surface of Landino's model into 1,700 miles, for example, so that he can divide it easily into ten parts of 100 miles each and ten parts of seventy miles each, reflecting the structure down below as he had slightly redesigned it.

11. Mathematics Old and New

1. Mario Biagioli, "The Social Status of Italian Mathematicians 1450–1600," *History of Science* 27 (1989): 41–95.

2. Niccolò Tartaglia, *General trattato di numeri et misure* (Venice: Per Curtio Troiano de i Nauó, 1556–1560).

3. Dante, *Paradiso* XXVII, 142–143.

4. Christopher Clavius, *Opera Mathematica*, v. 5 (Moguntiae: sumptibus A. Hierat, excudebat Reinhardvs Eltz, 1611–1612).

5. Ioannis Regiomontanus, *De triangulis planis et sphaericis libri quinque, una cum tabulus sinuum* . . . , 1464, ed. and repr. by Daniel Santbech, in *Problematum astronomicorum et geometricorum* . . . (Basileae: per Henrichum Petri, 1561). For extensive information on Regiomontanus see Paul Lawrence Rose, *The Italian Renaissance of Mathematics* (Geneva: Libraire Droz, 1975), 90–117.

6. Glen Van Brummelen, *The Mathematics of the Heavens and the Earth* (Princeton, NJ: Princeton University Press, 2009).

7. Ibid., 149.

8. Ptolemy actually computed the chord, a functional equivalent of the sine.

9. Ioannis Regiomontanus, *Epitoma in Almagestum Ptolomaei* (Venice: Johannes Hamman, 1496).

10. Santbech, *Problemata astronomicorum et geometricorum.*

11. The word seems to have been coined by Regiomontanus.

12. *coelum terrae misceant* . . . , Santbech, *Problemata astronomicorum et geometricorum,* 10.

13. *Difficile, me hercle, est huic hominum generi omnino os occludere et pestiferam calumniandi licentiam e medio tollere* . . . , ibid.

14. Ibid., 210.

15. One picture does show the cannonball dropping a little bit at discrete odd moments, making a zigzag line that cannot be intended as realistic.

16. N. Tartaglia, *Nova Scientia* (Venice: Stephano da Sabio, 1537). Tartaglia rounds the acute angle at the top of the trajectory with the arc of a circle.

17. The *Epitome* actually preceded *Almagest* into print.

18. For a short biography of Guidobaldo, see Rose, *The Italian Renaissance of Mathematics,* 222–242.

19. Jürgen Renn, Peter Damerow, Simone Rieger, and Michele Camerota, "Hunting the White Elephant—When and How Did Galileo Discover the Law of Fall?" available at www.mpiwg-berlin.mpg.de/Preprints/P97.PDF.

20. Guidobaldo to Galileo, December 1590, *Galileo Opere* X, 46. Guidobaldo shared Galileo's letters with his own son Horatio, who looked for them eagerly.

21. *Galileo Opere* I, 215–220.

22. Renn et al., *Hunting the White Elephant*, 62.

23. *Galileo Opere* I, 225–228.

24. Renn et al., *Hunting the White Elephant*, 72.

25. Guidobaldo del Monte to Galileo, 21 February 1592, *Galileo Opere* X, 47, quoted in Renn et al., *Hunting the White Elephant*, 59. In the same letter Guidobaldo expresses condolences on the death of Galileo's father.

26. Renn et al., *Hunting the White Elephant*, 75.

27. Ibid., 65.

28. Guidobaldo del Monte to Galileo, 10 January 1593, *Galileo Opere* X, 54.

29. Renn et al., *Hunting the White Elephant*, 41–45.

30. Ibid., 60–61.

12. Transforming Mathematics

1. Lucio Russo, *The Forgotten Revolution*, tr. and collab. Silvio Levy (Berlin: Springer Verlag, 2004).

2. V. Galilei, *Dialogue on Ancient and Modern Music*, tr. Claude Palisca (New Haven, CT: Yale University Press, 2003), 6.

3. Guidobaldo del Monte had said this explicitly in his *Liber Mechanicorum* (1577), 243–244, cited in Paul Lawrence Rose, *The Italian Renaissance of Mathematics* (Geneva: Libraire Droz, 1975), 232.

4. Daniel Santbech, *Problematum astronomicorum et geometricorum* . . . (Basileae: per Henrichum Petri, 1561), 10.

5. Kepler rederives a theorem of Pappus on volumes of revolution. It is odd that he seems not to know the reference, because he cites Pappus in another context in the same book.

6. Alexandre Koyrè, *Galileo Studies*, tr. John Mepham (Atlantic Highlands, NJ: Humanities Press, 1978), 37.

7. T. Settle, "An Experiment in the History of Science," *Science* 133 (1961): 19–23.

8. *Galileo Opere* I, 225–228.

9. Galileo, *Dialogue Concerning the Two Chief World Systems*, tr. and ed. Stillman Drake, foreword by Albert Einstein, 2nd rev. ed. (Berkeley: University of California Press, 1967), 282.

10. Ibid., 281.

11. Galileo, *Two New Sciences,* tr. Henry Crew and Alfonso de Salvio with an introduction by Antonio Favaro (New York: Dover Publications, 1954), 178.

12. Horst Bredekamp, *Galilei der Künstler: Der Mond, die Sonne, die Hand* (Berlin: Akademie Verlag, 2009).

13. Pierre Duhem, Letter to Father Bulliot, in Duhem, *Essays in the History and Philosophy of Science,* tr. and ed. Roger Ariew and Peter Barker (Indianapolis: Hackett, 1996), 161.

14. Annaliese Maier, *An der Grenze von Scholastik und Naturwissenschaft* (Essen: Essener Verlaganstalt, 1943), 331ff.

15. J. Renn, P. Damerow, and S. Rieger, "Hunting the White Elephant," in *Galileo in Context,* ed. Jürgen Renn (Cambridge: Cambridge University Press, 2000), 29–149.

16. Galileo to Castelli, 3 December 1639, *Galileo Opere* XVIII, 125–126.

17. Niccolò Aggiunti, *Oratio in mathematicae laudibus* (Rome: Mascardi, 1627), 5, translated by Philippa Goold.

13. The Oration

1. Galileo to Kepler, 28 August 1627, *Galileo Opere* XIII, 375.

2. Aggiunti, *Oratio de mathematicae laudibus* (Rome: Mascardi, 1627), 3. All translated passages of this work are by Philippa Goold.

3. Ibid., 7–8.

4. Antonio Favaro calls him "a born disciple of Galileo."

5. Giovanni Ciampoli, a friend and admirer of Galileo, was confidential secretary to Pope Urban VIII. For his support of Galileo, Ciampoli lost his high position in the Galileo Affair.

6. Ciampoli to Galileo, 10 July 1627, *Galileo Opere* XIII, 364.

7. Stillman Drake and C. D. O'Malley, *The Controversy on the Comets of 1618: Galileo Galilei, Horatio Grassi, Mario Guiducci, and Johannes Kepler* (Philadelphia: University of Pennsylvania Press, 1960).

8. It is a noteworthy coincidence that the *Oratio* and Galileo's recent book *The Assayer* (1623) were given to the same Roman printer, G. Mascardi.

9. Aggiunti to Galileo, 23 December 1626, *Galileo Opere* XIII, 344.

10. *Galileo Opere* II, 367.

11. *Orazione di Marcantonio Pieralli Rettore del Collegio della Sapienza di Pisa, recitata pubblicamente da lui l'Anno 1638. nella medesima Sapienza, in memoria di Niccolò Aggiunti Professor di Matematica nello Studio Pisano,* in Targioni-Tozzetti, *Notizie degli aggrandimenti delle scienze fisiche accaduti in Toscana nel corso di anni LX del secolo XVII* (Florence: Giuseppe Tofani, 1780), vol. 2, part 1, 259–274.

12. The phrase is similar to the slightly odd one quoted above from Gherardini, "one could travel securely, without hindrance, through heaven and earth." Perhaps Gherardini left out "in thought."

13. Aggiunti, *Oratio*, 8.

14. Galileo, in *Discoveries and Opinions of Galileo*, tr. and ed. Stillman Drake (New York: Doubleday Anchor Books, 1957), 237.

15. Aggiunti, *Oratio*, 9.

16. The shipbuilding is described in detail worthy of someone who knows his way around the Venetian Arsenal: the inventiveness of mathematics showed them "how to build ships and construct the hollow of the hold from connected beams; to build up the sides of the ship with pine planks like ribs; glue the seams, like joints, with tow caulking smeared with pitch; and finally fit the ship with oars, sails, and a rudder like a rein for the wooden horse on which, as Plautus says, they could ride over the blue roads."

17. Among a great many clocks enumerated here the only one that the Orator describes in detail is a clock whose little gears are driven by weights that chimes the intervals of the hours with little hammers. Galileo in these years was much occupied in trying to maintain such a clock for his daughter Sister Maria Celeste. When it worked, the clock summoned the nuns to prayer at appropriate hours of the night.

18. Aggiunti, *Oratio*, 9–16.

19. That Galileo should have had artistic associations with his astronomical discoveries is amply documented in Horst Bredekamp, *Galilei der Künstler: Der Mond, die Sonne, die Hand* (Berlin: Akademie Verlag, 2007).

20. Aggiunti, *Oratio*, 21.

21. Reproduced in Drake, *Galileo at Work*, 290.

22. Not easy, but also not impossible. In the summer of 1624 Galileo replied to the entreaties of a Genovese admirer, Bartolomeo Imperiali, with a microscope and instructions for using it. Imperiali and his circle were delighted with it, and in gratitude Imperiali sent Galileo a diamond ring.

23. Harmatian mode, or "nome," is obscure. It is the mode that roused Alexander the Great to arms according to Plutarch's *Second Oration Concerning the Fortune or Virtue of Alexander the Great*, an incident cited by V. Galilei in his *Dialogue on Ancient and Modern Music.*

24. This phrase closely echoes what is sometimes called Galileo's "mathematical manifesto" in *The Assayer*, speaking of the book of nature, "It is written in the language of mathematics, and its characters are triangles, circles, and other geometric figures."

25. Aggiunti, *Oratio,* 27–28. Ixion meant to ravish Juno, but Zeus contrived the cloud.

26. The first step in preparing a fleece, i.e., remedial teaching.

27. Aggiunti, *Oratio,* 28.

28. Ibid., 32.

29. Galileo, *Two New Sciences,* tr. Henry Crew and Alfonso de Salvio with an introduction by Antonio Favaro (New York: Dover, 1954), 98–107.

30. Sopra le definizioni delle proporzioni d'Euclide: principio di giornata aggiunta ai Discorsi e dimostrazioni matematiche intorno a due nuove scienze (giornata quinta), *Galileo Opere* VIII, 347–362.

Epilogue

1. A. Favaro, *Amici e Corrispondenti* (Florence: Libreria Editrice Salimbeni, 1983), v. 2, pp. 1042–1043.

2. See appendices in A. Favaro, *Amici e Corrispondenti,* v. 3, pp. 1221–1243.

3. Kepler to Magini, 10 May 1610, *Galileo Opere,* 353.

4. Kepler to Herwart von Hohenberg, 26 March 1598, *Galileo Opere* X, 72.

5. Galileo, *Dialogo,* in *Galileo Opere* VII, 486.

6. Galileo to Micanzio, 7 November 1637, *Galileo Opere* XVII, 214–215.

7. Galileo to Rinuccini, 29 March 1641, *Galileo Opere* XVIII, 314–315, tr. Stillman Drake, in *Galileo at Work* (Mineola, NY: Dover, 1978), 417.

8. Indeed, Galileo's other favorite argument for Copernicanism, occupying much of the Second Day of the *Dialogue,* was the principle of relativity: motion in common is as if it did not exist, as he put it. That idea had always been somewhat in tension with his theory of the tides, to be sure, but now he had given up the tidal theory. What remained, especially if he took the principle of relativity seriously, might have persuaded him that there was no objective meaning to saying that the Earth moved.

9. John Milton visited Galileo in the summer of 1638. It is not known what they talked about, but it is not inconceivable that, in the words of Raphael,

> To ask or search I blame thee not, for Heav'n
> Is as the Book of God before thee set,
> Wherein to read his wondrous Works, and learne
> His Seasons, Hours, or Days, or Months, or Yeares:
> This to attain, whether Heav'n move or Earth,
> Imports not, if thou reck'n right, the rest

From Man or Angel the great Architect
Did wisely to conceal, and not divulge
His secrets to be scann'd by them who ought
Rather admire; or if they list to try
Conjecture, he his Fabric of the Heav'ns
Hath left to thir disputes, perhaps to move
His laughter at thir quaint Opinions wide.

It is not inconceivable, I say, that in *Paradise Lost* Milton recorded Galileo's ultimate opinion on the subject.

Acknowledgments

I wish to thank Mount Holyoke College for its steady support of this work, in ways both tangible and intangible. The idea for a book occurred to me in Mathematics Across the Curriculum, a faculty seminar jointly sponsored by the National Science Foundation and the National Endowment for the Humanities and organized by Giuliana Davidoff, Robert Schwartz, and Donal O'Shea. Don's subsequent support of the project as Dean of Faculty has been characteristically unstinting. Angelo Mazzocco and I, as partners in the seminar, began to investigate what mathematics meant in the Renaissance, a question that came to occupy me increasingly over the years. Members of the Classics Department were generous in their help when I had questions about Latin. In particular the contributions of Philippa Goold to the last chapter leave me with a debt that I can never adequately repay. Her translation opened the *Oratio* for me beyond anything that my own abilities could have managed, and I treasure the memories of afternoons with her going over the details of that fascinating text. Peter Pesic gave me thoughtful critical comments on many different drafts, and I am also indebted to anonymous readers on crucial points. I benefited from trying out these ideas on various audiences, thanks to speaking invitations from Peter Pesic, Arielle Saiber, Giuseppe Mazzotta,

Acknowledgments

Arthur Kinney, and Massimo Ciavolella. Editors Chandler Davis, Jan Tobochnik, and Harvey Gould let me express some of these ideas in print, and Michael Fisher at Harvard University Press agreed to give them all a unified forum in this volume. Wendy Watson helped me with permissions. I am grateful to libraries at Mount Holyoke, Columbia, and Harvard, and to all those at European Cultural Heritage Online (ECHO) and the Istituto e Museo di Storia della Scienza (IMSS —now Museo Galileo) who have made rare books freely available online. In the last analysis my most constructive critics have been my own family, Maya and Indira.

Index